# Ionic Equilibrium

*The Author*

**Dr L. K. Mishra** obtained his Post-Graduate and Doctorate degree in Biochemistry from Narendra Deva University of Agriculture and Technology, Faizabad, Uttar Pradesh. He started his academic career as Assistant Professor in Biochemistry at ND University of Agriculture and Technology Faizabad. He later served as Assistant Professor at V.B.S. Purvanchal University Jaunpur, U.P. Presently he is working as Assistant Professor (Biochemistry) at Central Agricultural University, Imphal, Manipur. He is involved in teaching Biochemistry for the last fourteen years. The author has also handled research projects funded by different agencies of the Government of India. The focus of these projects is metabolic profiling of the indigenous food products of the North East India and development of new products utilizing the biodiversity of the NE India. He has authored several research papers and review articles in International and National Journals of repute and is associated with many national and international Societies in the field of Plant Biochemistry. He worked as DBT Overseas Visiting Associate at Department of Plant Sciences, North Dakota State University, Fargo, United States. The focus of the research at North Dakota State University was exploration of phenolic linked antioxidant activities of food systems for health promotion and combating metabolic syndromes like type-2 diabetes, hypertension. In recognition of his work in the field of medicinal plants and agricultural research and development he has been awarded fellowship for the year 2019 by International Society of Noni Science, Chennai.

# Ionic Equilibrium

*Author*

**Dr L. K. Mishra**

Kruger Brentt
Publishers
2020

Kruger Brentt Publishers UK. LTD.
Company Number 9728962

Regd. Office: 68 St Margarets Road, Edgware, Middlesex HA8 9UU

---



Library of Congress Cataloging-in-Publication Data

---

ISBN 978-1-78715-012-6 (Hardbound)

---

For information on all our publications visit our website at http://krugerbrentt.com/

# Preface

Ionic equilibrium is a state in which the rate of the forward response equates to the rate of the backward response. Typically, this state results when the forward response profits at the very same rate as the reverse response. The response rates of the forward and backward responses are typically not absolutely no, however equivalent. As the response items collect, the concentration of each reactant reduces and so does the response rate. Ionic equilibrium is a state where the rate of the forward response amounts to the rate of the backward response. To puts it simply, there is no net modification in concentrations of items and reactants. This sort of equilibrium is likewise called vibrant equilibrium.

The significance of ionic equilibrium is the point at which the concentrations of products and reactants do not modify with time. It looks like though the response has actually stopped however in truth the rates of the reverse and forward responses are equivalent so products and reactants are being produced at the precise very same rate. At equilibrium, the response can lie far to the right, showing that there are more products around at equilibrium, or far to the left, recommending that at equilibrium there are likewise reactants. The concentration of the reactants and outcomes in a reaction at equilibrium can be revealed by equilibrium continuous, represented K or Keg.

The present book entitled *"**Ionic Equilibrium**"* describes about equilibrium, equilibrium constant, ionic strength and activity coefficients, strong acids and bases, monoprotic and polyprotic acids and bases, solubility, complex organic compounds, oxidation-reduction equilibria, carbon dioxide, pH and equilibrium. It also explains how various analytical methods help to calculate the different species that result from equilibria in aqueous solutions and it is supplemented with numerous examples, diagrams and tables for easy understanding and better retention.

It is an ideal course text for students of engineering, environmental chemistry, analytical chemistry, geochemistry, biochemistry, physical chemistry, clinical chemistry and plant chemistry. Constructive suggestions of readers especially the teaching faculty and students are invited for further improvement of the book. Their esteemed suggestions will be duly acknowledged in the future editions.

I am highly thankful to Publishers Kruger Brentt Publisher, London (UK) for his interest shown in publishing of the book so efficiently and promptly.

***Dr L. K. Mishra***

# Contents

# 1 Equilibrium

## 1.1 System of Equilibrium

### 1.1.1 Equilibrium

Equilibrium is a state of chemical reaction where future changes in composition of reactant and the product mixtures cannot be alleged from external point of view. Further, analyzing that inside the mixture what idea will result out without absence of actual change or whether it takes place at an equal rate from either side making it seem that real change doesn't occur.

### 1.1.2 Chemical Equilibrium

Chemical equilibrium describes that the concentrations of reactants and products in a reaction which takes place in a closed system, which is no longer a change with respect to time. Also, it is known that the rate of forward reaction will be equal to the rate of reverse reaction, where the concentrations of reactants and products remain quiet stable in a chemical reaction.

Such type of stable pattern in a system is normally known as equilibrium. An equilibrium is said to be stable, if the system always returns to it after small disturbances. If the system moves away from the equilibrium after small disturbances, then the equilibrium is unstable (Figure 1.1).

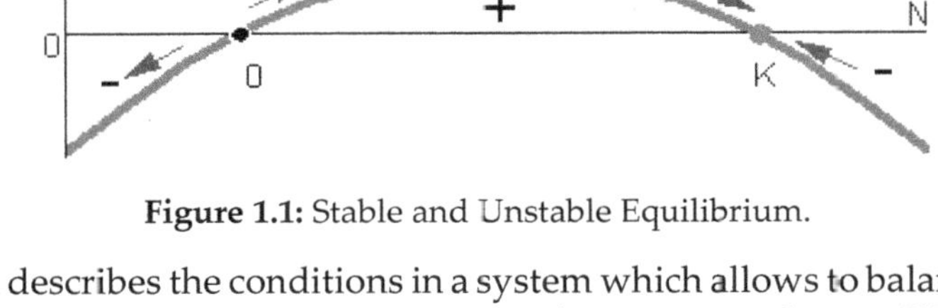


**Figure 1.1:** Stable and Unstable Equilibrium.

Equilibrium describes the conditions in a system which allows to balance, resulting in no net change. There can be two ways to achieve no net change. When a system is at rest, it is known as static equilibrium while a system achieving a balance between equal and opposite forces is considered as dynamic equilibrium.

Consider the reaction occurring in a closed container:

$$H_2 + I_2 \rightarrow 2HI$$

It is simply a reaction which occurs among elemental hydrogen and elemental iodine for making hydrogen iodide. Here, all the $H_2$ and $I_2$ molecules on reacting will make HI, but a reverse chemical reaction is also taking place as:

$$2HI \rightarrow H_2 + I_2$$

The equation act to undo what the first reaction does. Eventually, the reverse reaction proceeds so quickly that it matches the speed of the forward reaction. At that time, any continued overall reaction will stops which makes the reaction to reached at chemical equilibrium which is the point where forward and reverse processes balance each other's.

As two opposing processes takes place at once, so it is conventional to represent an equilibrium using double arrow as:

$$H_2 + I_2 \rightleftharpoons 2HI$$

The double arrow shows that the reaction is going in both directions which should be balanced.

When the concentrations of reactants and products have become constant, an equation is said to have reached a point of equilibrium. The consistency of measurable properties such as concentration, color, pressure and density can show a state of equilibrium. The equilibrium state is said to be dynamic, meaning that the reaction is continuously in motion. This consistency, however, does not mean that the reactions have stopped, but rather that the rates of the two opposing reactions have become equal. The amount of products and reactants produced are consistent, and there is no net change.

There are many examples of chemical equilibrium all around you. One example is a bottle of cool drink. In the bottle there is carbon dioxide ($CO_2$) dissolved in the liquid. There is also $CO_2$ gas in the space between the liquid and the bottle cap. There results constant movement of $CO_2$ from the liquid to gas phase and from gas phase to liquid phase. It is observed that there will no change in the bottle of the cold drink, hence the system is in equilibrium and results as:

$$CO_2(g) + H_2O(\ell) \rightleftharpoons H_2CO_3(aq)$$

Consider another example of chemical equilibrium of a Haemoglobin which is a macromolecule that transports oxygen around our bodies. Without it we would not survive. The haemoglobin has to be able to take up oxygen, but also to release it and this is done through changes in the chemical equilibrium of this reaction in different places in our bodies.

$$\text{Haemoglobin}\ (aq) + 4O_2(g) \rightleftharpoons \text{haemoglobin}\ (O_2)\ 4(aq)$$

Figure 1.2 shows the graphical representation of chemical equilibrium. In this, starting from the beginning of reaction shown by y-axis or at time = 0, the rate of forward reaction will tends to rise sharply while rate of reverse reaction declines which happens as reaction consisting of pure reactants.

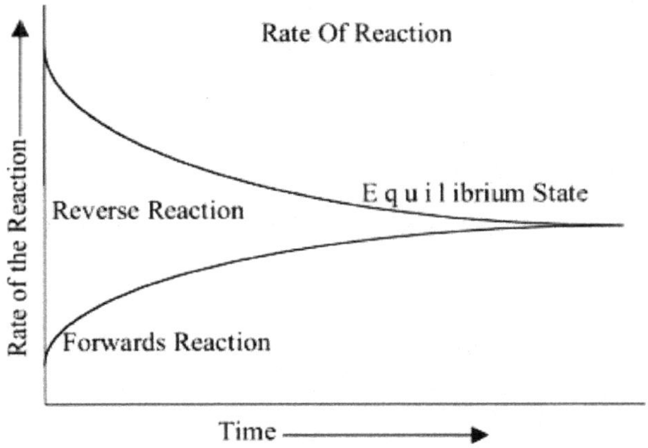


**Figure 1.2:** Graphical Representation of Chemical Equilibrium.

For an advance reaction, reactants are converted to products till a large concentration of products are available, where a reverse reaction becomes a factor and at this point it results in equilibrium, where forward and reverse rates converge at same point forming an equilibrium state.

## 1.1.3 Characteristics of Chemical Equilibrium

### Stability of Observable Properties

It is observed that state of chemical equilibrium is characterized by constancy of certain properties which includes:

- Concentration
- Density
- Pressure
- Color.

On the other hand, the state of equilibrium is observed only as long as the reaction conditions like temperature and pressure are kept fixed.

### Fastened by a Catalyst

Catalyst does not affect the equilibrium of a system, but helps in speeding up a reaction. It will speed up the forward as well as reverse reaction which helps the reaction in reaching its equilibrium faster. It is observed that similar amount of reactants and products are present at equilibrium in a catalyzed or non-catalyzed reaction. Catalyst only facilitates the reaction to proceed through a lower-energy transition state of reactants to products.

In an example,

$$2SO_2+O_2 \leftrightarrow 2SO_3$$

it is noted that the above reaction takes place at 500 degree which takes longer time to reach in equilibrium. On the other hand, time needed for attaining the equilibrium will be reduced by adding platinum or vanadium pentoxide. In this, catalyst helps the equilibrium in attaining faster rate without any changes in composition of reaction mixture i.e. reactants $SO_2$, $O_2$ and the product $SO_3$.

**Chemical Equilibrium can be Attained from Either Direction**

In an equilibrium stage, many physical states co-existed under a given set of conditions and that there was no change in measurable properties of the system. Chemical change involves interaction among reactants to form the products. The state where reactants and products co-exist without further chemical changes is a chemical equilibrium where chemical reaction takes place when it is a reversible reaction.

With this, it is possible to have a state of equilibrium in a reverse direction by considering actual products as reactants. For a reaction, it can be

$$H_2 + I_2 \leftrightarrow 2HI$$

OR

$$2HI \leftrightarrow H_2 + I_2$$

Here, the equilibrium state is similar in both the reactions. At equilibrium, gases $H_2$, $I_2$, 2HI will be present in similar quantity whether the reaction forms hydrogen iodide or decomposition of hydrogen iodide.

**Chemical Equilibrium is Dynamic in Nature**

Chemical equilibrium is an example of a dynamic equilibrium and not static equilibrium. Know the difference in the two. Static equilibrium is fixed and non-changing – like balancing weights on a balance beam. Dynamic equilibrium has no NET overall change but does have some given processes still proceeding. The process itself proceeds both forwards and backwards at exactly the same rate. Stated chemically, equilibrium is achieved when the forward rate of reaction equals the reverse rate of reaction.

Constancy of observable property in an equilibrium will not relate to stopping of reaction, but after attaining equilibrium, the reaction does not stop as it appears to have been stopped. Since the rates of two opposing reactions become equal, so if products are formed from the reactants, then exactly same amount of reactants will be obtained from the products.

In a reaction between $H_2$ and $I_2$ to form HI, at equilibrium, when certain amount of hydrogen iodide is formed by combination of hydrogen and iodine, then same amount of hydrogen iodide gets decomposed resulting in $H_2$ and $I_2$

$$H_2(g) + I_2(g) \rightleftharpoons 1\ HI_{(g)}$$

Here, rate of formation of HI at equilibrium will be equal to the rate of decomposition of HI where the concentrations of reactants and products remains constant. Both the opposing reactions will continue at similar rate even after the equilibrium, so such type of chemical equilibrium is known as dynamic equilibrium.

### 1.1.4. Static Equilibrium

Static equilibrium is a type of equilibrium where rates of forward and reverse processes are zero and that occurs when there is no exchange between reactants and products. In static equilibrium, the number of reactants and products remain the same. On the other hand, reactions itself has come to halt without any extra reactants for converting into products and vice versa which shows static condition.

For example, if a reaction is started with several reactant molecules and few product molecules, the after some time, reaction will get halt thereby keeping the composition same without any interchanges in mixture. An example of static equilibrium is diamond turning into graphite.

$$C\ (\text{diamond}) \rightleftharpoons C\ (\text{graphite})$$

Graphite is more stable than diamond, but the activation energy is so high that it should be heated to about 2000°C before you'll see any diamond converting to graphite.

### 1.1.5 Dynamic Equilibrium

Dynamic equilibrium is a reversible chemical process where forward and reverse rates of reaction are similar. Dynamic equilibrium is a state where nothing happens and there results no net change, but is also a state where chemical reactions are taking place at rapid rates. Such type of reactions are normally opposites of each other so that the net change will be nothing. Any chemical system at equilibrium is a dynamic system where real reactions are occurring with real speed and allows such system to be altered in certain way through external means, so that the reactions can operate to bring it back to equilibrium again.

#### Static Vs. Dynamic Equilibrium

In static equilibrium, microscopic and macroscopic properties remain unaltered while in dynamic equilibrium, microscopic properties change but macroscopic properties remains unchanged.

Also in a chemical system, if the rate of forward reaction and the backward reaction are the same, the system is said to be in dynamic equilibrium.

### 1.1.6 Open and Closed System

#### Open System at Equilibrium

The chemical reaction that happens in a container where one or more of reactants or products can able to escape is known as an open system. In a closed system, equilibrium is achieved when opposing variables in the system are balance. Here, the equilibrium results as static or dynamic where static is commonly found in closed systems while dynamic is a property of an open system.

Living systems are open systems where recurrent alteration of fluxes of matter, energy and information with their equilibrium is dynamic. It is observed that dynamic equilibrium is known as flux equilibria or steady state that can be used interchangeably in both closed and open systems.

An important quality of open system is that it can perform work, which is unachievable in a closed system in an equilibrium state as closed system in equilibrium will not require energy for the preservation of its state nor can energy be obtained from it. For doing work, it is necessary that an open system is not in an equilibrium state. On the other hand, system has a tendency to attain such a state as a result of which, a equilibrium found in any open system is not a true equilibrium as it is incapable of performing work. Instead, it is a dynamic pseudo-equilibrium which is kept constant at certain distance from true equilibrium.

**Behaviour of Open Systems**

It is necessary for many systems to maintain their equilibrium in changing environments or disturbances, otherwise they cannot function properly or their goals cannot be attained. In living systems, the process of self-maintenance or 'homeostasis' is essential to ensure their survival and viability. With homeostasis, which results as a process through which a system preserves its existence by maintenance of its dynamic equilibrium which is further known as homeostatic equilibrium. Thus, a mature organism as an open system appears to be unchanged over a period of time because there is a continuous exchange and replacement of matter, energy, and information between the system and the environment.

**Closed System at Equilibrium**

If a chemical reaction happens in a container where none of reactants or products can escape, then such type of system is known as a closed system. Reversible reactions that happen in a closed system eventually reaches equilibrium.

Closed systems are useful for performing chemical reactions and in rearranging atoms inside a matter to give new combinations of atoms. Such systems are used in the laboratory as well as chemical industries such as in manufacturing of medicines where a reactions occur in a special vessel known as chemical reactor that has been sealed to prevent any matter from entering. The sealed reactor is now a closed system allowing for the careful control of the flow of energy across the walls of the reactor through heating or cooling.

However, many reactions, especially those involving gases, generally do not go to completion. In reverse reactions occur before the limiting reactants get completely used up. All gaseous reactions in a closed system undergo reversible processes and the following are some examples:

1. $N_2(g) + 3H_2(g) \rightleftharpoons 2\ NH_3(g)$;
2. $2SO_2(g) + O_2(g) \rightleftharpoons 2\ SO_3(g)$;
3. $CO(g) + H_2O(g) \rightleftharpoons CO_2(g) + H_2(g)$;

In aqueous solutions, the dissociations of weak acids and bases and the dissolution of slightly soluble salts are reversible processes.

1. $HC_2H_3O_2(aq) + H_2O(l) \rightleftharpoons H_3O^+(aq) + C_2H_3O_2^-(aq)$;
2. $NH_3(aq) + H_2O(l) \rightleftharpoons NH_4^+(aq) + OH^-(aq)$;
3. $Mg(OH)_2(s) \rightleftharpoons Mg^{2+}(aq) + 2\ OH^-(aq)$;

Consider the following reaction that occurs in a closed container:

$$2\ NO_2(g) \rightleftharpoons N_2O_4(g)$$

If an amount of $NO_2$ gas is placed in a sealed container and let it stands at room temperature, the brownish color of the gas slowly fades as the forward reaction that changes some of $NO_2$ (brown) to $N_2O_4$ (colorless) takes place. The fading of the color may continue for some times as more and more of $NO_2$ gas is converted to $N_2O_4$, but this change apparently seems to stop at some point when the reaction has reached a state of equilibrium. Equilibrium occurs due to reversible and processes occur at the same rate.

All chemical reactions done in a closed system are essentially reversible and will eventually reach to a state of equilibrium where, concentrations of components in reaction mixture remain unchanged even when the mixture is left for longer period at constant temperature.

Many chemical processes in nature also occur in reversible manner and established equilibrium. For example, the formation of stalactite and stalagmite in limestone caves are the results of reversible reactions. When rain water saturated with dissolved $CO_2$ seeps through limestone hills (composed mainly of $CaCO_3$), the following reaction occurs in which some of $CaCO_3$ (limestone) are dissolved.

$$CaCO_3(s) + H_2O(l) + CO_2(aq) \rightleftharpoons Ca^{2+}(aq) + 2\ HCO_3^-(aq);$$

The rain water that sieve through the rocks becomes saturated with $Ca^{2+}$ and $HCO_3^-$ ions causing the reverse reaction to occur, depositing $CaCO_3$ that eventually form the *stalactites* and *stalagmites*.

$$Ca^{2+}(aq) + 2HCO_3^-(aq) \rightleftharpoons CaCO_3(s) + H_2O(l) + CO_2(g);$$

The formation of stalactite and stalagmite in limestone caves is the result of following reversible reactions:

$$CaCO_3(s) + H_2O(l) + CO_2(aq) \rightleftharpoons Ca^{2+}(aq) + 2\ HCO_3^-(aq);$$

Some reactions occur at a significantly fast rate and reach equilibrium within minutes. Precipitation and acid-base reactions are examples of reactions that quickly reach equilibrium. Other reactions occur very slowly and may take weeks if not months or years to reach equilibrium.

The following reactions are examples of these slow reactions. They are usually carried out at high temperature and in the presence of catalysts:

$$N_2(g) + 3\ H_2(g) \rightleftharpoons 2\ NH_3(g);$$

$$2\ SO_2(g) + O_2(g) \rightleftharpoons 2\ SO_3(g)$$

$$CH_4(g) + H_2O(g) \rightleftharpoons CO(g) + 3H_2(g)$$

### 1.1.7 Factors Affecting Equilibrium

#### Changing Concentration

Consider a reversible reaction in a closed system and state of equilibrium as:

$$A(aq) + B(\mathrm{aq}) \rightleftharpoons C(aq) + D(aq)$$

On increasing the concentration of substance A on left-hand side, the rate of the forward reaction will increase. This shows that more of the products C and D are formed. With this, the rate of the reverse reaction starts increasing. Eventually, the rates of the forward and reverse reactions will be equal again and equilibrium is re-established. However, in response to the increased amount of A, it is noted that there will be more of C and D in the mixture as from the original.

#### Changing Pressure

In reactions involving gases, changing pressure will affect the composition of an equilibrium mixture. On applying Le Chatelier's principle, if pressure increases, equilibrium will shift in such a way so as to reduce the pressure.

$$X(g) + Y(g) \rightleftharpoons Z(g)$$

In an example above, each molecule of gas in the mixture will contribute to pressure inside the container. On reducing the number of gas molecules in a container, the pressure will decrease. If the position of equilibrium shifts to the right, then there will be low amount of gas molecules in the mixture which finally reduces the pressure.

#### Changing Temperature

It is known that on increasing the temperature of a mixture at equilibrium, the equilibrium will tends to shift. As per Le Chatelier's principle, the equilibrium position which will shift will try to reduce the temperature.

$$A(aq) + B(aq) \rightleftharpoons C(aq) + D(aq)\ \Delta H = -100\ \text{kJ/mole}$$

In a reaction, the forward reaction is exothermic which gives heat. If the forward reaction increases more than reverse reaction, more and more heat will be released. However, the reverse reaction is endothermic ($\Delta H = +100$ kJ/mole), taking in energy from its surroundings. Hence, increasing the rate of the reverse reaction will result in more energy absorption, which finally decreases the temperature. As a result of this, there will be more quantity of A and B in the new equilibrium mixture.

**Adding a Catalyst**

A catalyst has no effect on the composition of an equilibrium mixture. This is because it increases the rate of both the forward and reverse reactions to the same extent. However, catalysts are important in reversible reactions as because they increase the rate of reaction and saves time. With this, there will be more product in a certain time, even though the mixture, if allowed to proceed to a state of equilibrium, still have similar proportion of product in it.

In fact, the reversible reactions in industry usually take place in continuous processes in which the product is continuously removed and a state of equilibrium is never actually reached.

**Position of Equilibrium**

There are many equilibrium mixtures possible for a given chemical reaction, depending on the concentrations of solutions, the pressure of gases and the temperature of the reaction. The term position of equilibrium can be used to describe a particular set of equilibrium concentrations for a reaction. If either the temperature, pressure or concentration of one of the substances changes, then the concentrations of all the substances will change until a new position of equilibrium is reached.

Although the rates of the forward reaction and the reverse reaction are at equilibrium, the concentrations of the reactants and the concentrations of the products are rarely the same.

If the forward reaction is nearly complete before the rate of the reverse reaction has increased sufficiently to establish equilibrium, then the position of equilibrium lies towards the right.

If a low amount of the reactants have been converted to the products before equilibrium is reached, then the position of equilibrium lies to the left hand side. When a reaction reaches equilibrium, it is impossible to tell whether the reaction has started with the reactants or the products.

In an example, on reacting equal amounts of Hydrogen and Iodine to form Hydrogen Iodide, there results similar amount of reactants and products when decomposition of Hydrogen Iodide takes place.

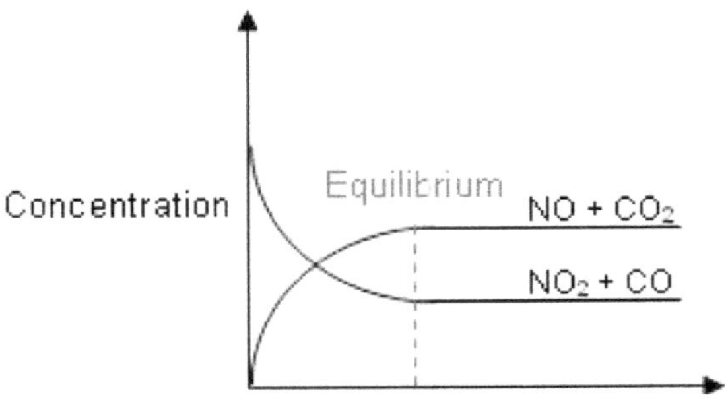

### 1.1.8 Reversible and Irreversible Equilibrium Reactions

Depending on the extent of reaction, reactions are of two types:

- Reversible
- Irreversible

Reactants, written on the left hand side of a chemical equation, proceed in their reactions. The products are written on the right hand side of the chemical equation. This is indicated by the direction of the arrow in a chemical equation. This is a direct reaction.

For example when steam is passed over hot iron, hydrogen gas and ferroso-ferric oxide are produced.

$$3Fe + 4H_2O \longrightarrow Fe_3O_4 + 4H_2 \text{ --------[Direct reaction]}$$

In some chemical reactions, as soon as the products are formed they begin to react among themselves and actually a backward or reverse reaction starts.

$$Fe_3O_4 + 4H_2 \longrightarrow 3Fe + 4H_2O \text{ --------[Reverse reaction]}$$

It is interesting to note that if carried out in a open vessel, the reaction cannot go in the backward direction because, one of the products of the reaction, the hydrogen gas escapes. But if the reaction is carried out in a closed vessel, the reaction becomes reversible, because when hydrogen gas is passed over hot ferroso-ferric oxide, steam and iron are produced.

These two reactions may be represented by a single equation as

$$3Fe(s) + 4H_2O_{(g)} \rightleftharpoons Fe_3O_{4(s)} + 4H_{2(g)}$$

Thus a chemical reactions that proceeds in two directions is known as a reversible reaction.

**Example**

To repeat, a chemical reaction in which substances react together to produce resultants and the resultants in turn react with one another to produce the original substances is known as reversible chemical reaction. For example, decomposition followed by combination reactions of calcium carbonate.

$$CaCO_{3(S)} \underset{\text{Combination reaction}}{\overset{\text{Decomposition reaction}}{\rightleftharpoons}} CaO_{(S)} + CO_{2(g)}$$

When we heat calcium carbonate ($CaCO_3$) in a closed vessel the evolved carbon dioxide will not go out; it reacts with another product calcium oxide (CaO) to give the calcium carbonate ($CaCO_3$) in reverse. The direction of reaction is affected by like reaction temperature, concentration of reactants, stability of products, etc.

When the decomposition reaction is carried out in an open condition, evolved carbon dioxide gas escapes. Therefore, uniting of $CO_2$ gas with calcium oxide is not observed resulting in an irreversible reaction.

$$N_{2(g)} + 3H_{2(g)} \rightleftharpoons 2NH_{3(g)}$$

$$2SO_{2(g)} + O_{2(g)} \rightleftharpoons 2SO_{3(g)}$$

$$PCl_{5(g)} \rightleftharpoons PCl_{3(g)} + Cl_{2(g)}$$

$$NH_4OH \text{ .... } NH_4 + (Aq) + OH^{\ominus} (Aq)$$

Reversible reactions do not proceed to completion in any one direction. Usually the rate of the reverse reaction is slower than that of the forward reaction.

**Irreversible Reactions**

Reaction in which entire amount of the reactants is converted into products is termed as irreversible reaction. So, irreversible reactions are those reactions in which the products manufactured by reaction in reactants cannot be converted back in reactants. These reactions are represented by putting single headed arrow (→) between reactants and products.

The cooking gas used in the kitchen is butane, which on combustion with air produces carbon dioxide and water as products.

$$C_4H_{10}(g) + O_2 (g) \rightarrow CO_2(g) + H_2O (g) + \text{Heat}$$

In this reaction we cannot combine carbon dioxide and water to remanufacture butane so this is an irreversible reaction.

**Irreversible Reaction where Reverse Reaction occurs to Negligible Extent**

In this, reverse reaction occurs to such a small extent that it can be avoided. The only reaction that occurs is the conversion of reactants to product. So every reactants are converted to product. Now at equilibrium [A] = 0, as 1 mol of A reacted, which should form 1 mol of P and 1 mol Q - i.e. the concentration of products at equilibrium is 1 M.

At an earlier time of reaction, when A = 0.8 M, only part of reactants have reacted is 0.2 M which produces equal amount of products P and Q. Here, figures of [A] and [P] as a function of time is shown, where A decreases in nonlinear manner to 0 M while P increases in fractional manner to 1 M concentration as shown.

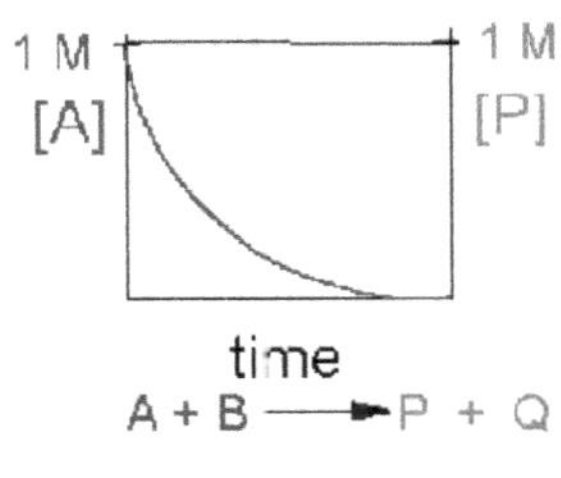

Examples of irreversible reactions are reactions of strong acids such as nitric, sulfuric, hydrochloric with bases such as $OH^-$ and water or it can combustion reactions such as burning of sugars and hydrocarbons forming carbon dioxide $CO_2$ and water $H_2O$.

**Characteristics**

a) These reactions proceed only in one direction (forward direction),

b) These reactions can proceed to completion,

c) In an irreversible reaction, $\Delta G < 0$,

d) The arrow ($\rightarrow$) is placed between reactants and products,

**Examples**

a) Neutralisation between strong acid and strong base e.g.,

$NaOH + HCl \rightarrow NaCl + H_2O$

b) Double decomposition reactions or precipitation reactions e.g.,

$BaCl_2 + H_2SO_4 \rightarrow BaSO_4$

c) Thermal decomposition e.g.,

$2KClO_3 \rightarrow 2KCl + 3O_2$

d) Redox reactions e.g.,

$SnCl_2 + 2FeCl_3 \rightarrow SnCl_4 + 2\ FeCl_2$

## 1.2 Multiple Equilibria

Multiple equilibria is a type of equilibrium reactions which on adding results in total reaction where an equilibrium constant for total reaction results as equal with the product *of equilibrium constants for contributing reactions.*

Mathematically, this can be written as:

$$A_1(g) + B_1(g) + \ldots \rightleftharpoons C_1(g) + D_1(g) + \ldots \qquad K_{c1}$$

$$A_2(g) + B_2(g) + \ldots \rightleftharpoons C_2(g) + D_2(g) + \ldots \qquad K_{c2}$$

$$A_3(g) + B_3(g) + \ldots \rightleftharpoons C_3(g) + D_3(g) + \ldots \qquad K_{c3}$$

etc.

$$A_1(g) + A_2(g) + A_3(g) + B_1(g) + B_2(g) + B_3(g) + \ldots \rightleftharpoons C_1(g) + C_2(g) + C_3(g) + D_1(g) + D_2(g) + D_3(g) + \ldots \qquad K_{ctotal} = K_{c1} \times K_{c2} \times K_{c3} \times \ldots$$

**Example**

Find the equilibrium constant for the reaction:

$2\ N_2O(g) + 3\ O_2(g) \rightleftharpoons 4\ NO_2(g)$ given the following information:

$N_2O(g) + ½\, O_2(g) \rightleftharpoons 2\, NO(g)$ $K_c = 1.7\times10^{-13}$

$NO(g) + ½\, O_2(g) \rightleftharpoons NO_2(g)$ $K_c = 6.83\times10^{6}$

**Solution**

The above problem is solved assuming it as multiple equilibria where reactions are used with the given equilibrium constants in order to have the reaction whose equilibrium constant is unknown. Here the target reaction is

$$2\, N_2O(g) + 3\, O_2(g) \rightleftharpoons 4\, NO_2(g)$$

In this, reaction (**1**) has correct reactants as with incorrect stoichiometry and wrong products, it results as:

$$N_2O(g) + ½\, O_2(g) \rightleftharpoons 2\, NO(g)$$

For having $N_2O$ correct stoichiometry, apply (1) again:

$$N_2O(g) + ½\, O_2(g) \rightleftharpoons 2\, NO(g)$$

Summing these two reactions gives:

$$2\, N_2O(g) + 1\, O_2(g) \rightleftharpoons 4\, NO(g)$$

Also, reaction (2) will be used to remove NO product which will not show in final target reaction. However, by using it four times in order to fully remove NO:

$$NO(g) + ½\, O_2(g) \rightleftharpoons NO_2(g)$$

$$NO(g) + ½\, O_2(g) \rightleftharpoons NO_2(g)$$

$$NO(g) + ½\, O_2(g) \rightleftharpoons NO_2(g)$$

$$NO(g) + ½\, O_2(g) \rightleftharpoons NO_2(g)$$

Hence sum of all these reactions will be:

$$2\, N_2O(g) + 3\, O_2(g) + 4\, NO(g) \rightleftharpoons 4\, NO_2(g) + 4\, NO(g)$$

The 4 $NO(g)$ can be eliminated from both sides of the reaction to give the desired reaction. Then, the equilibrium constant is found by multiplying the equilibrium constants from each contributing reaction:

$$K_c = K_{c1} \times K_{c1} \times K_{c2} \times K_{c2} \times K_{c2} \times K_{c2}$$

$$= (1.7\times10^{-13})(1.7\times10^{-13})(6.83\times10^{6})(6.83\times10^{6})(6.83\times10^{6})(6.83\times10^{6})$$

$$= 63$$

## 1.3 Concentration Scales

Concentration is a very common concept used in chemistry and related fields. It is the measure of how much of a given substance there is mixed with another substance. This can apply to any sort of chemical mixture, but most frequently is used in relation to solutions, where it refers to the amount of solute dissolved in a solvent. To concentrate a solution, one must add more solute, or reduce the amount of solvent. By contrast, to dilute a solution, one must add more solvent, or reduce the amount of solute. There exists a concentration at which no further solute will dissolve in a solution. At this point, the solution is said to be saturated. If additional solute is added to a saturated solution, it will not dissolve. Instead, phase separation will occur, leading to either coexisting phases or a suspension.

Concentration is an expression of how much solute is dissolved in solvent in a chemical solution. There are multiple units of concentration which tells you how the concentration depends on the use the chemical solution (Figure 1.3).

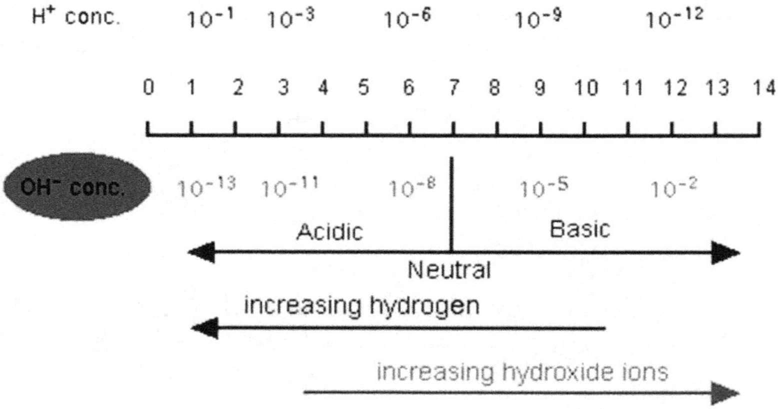


**Figure 1.3:** Concentration Scale.

A scale defining many numerical systems related to quantitative relation of components of a mixture such as for solutions where the concentration is expressed as mass, volume or number of moles of solute in proportion to solvent or total solution.

The effective strength of an acid or base in solution, or acidity, will vary with both the intrinsic strength of the acid or base itself and the amount of the acid or base which is present in the aqueous solution. As in the case of an aqueous solution of a strong acid, the contribution from the autoionization of water is almost always neglegible. In an aqueous solution of a strong base, acidity is essentially determined by the concentration of the strong base alone.

Quantitative notation of concentration is far more informative and useful from a scientific point of view. There are a number of different ways to quantitatively express concentration such as:

### 1.3.1 Mass Percentage

It shows mass of a substance in a mixture as a percentage of the mass of the entire mixture. For instance: if a bottle contains 40 grams of ethanol and 60 grams of water, then it contains 40% ethanol by mass. Commercial concentrated aqueous reagents such as acid and bases are often labeled in concentrations of weight percentage with the specific gravity also listed.

### 1.3.2. Mass-Volume Percentage % m/v or % w/v

It shows mass of a substance in a mixture as a percentage of the volume of the entire mixture. Mass-volume percentage is often used for solutions made from solid reagents. It is the mass of the solute in grams multiplied by one hundred divided by the volume of solution in milliliters.

Volume-volume percentage or % (v/v): describes the volume of the solute in mL per 100 mL of the resulting solution. This is most useful when a liquid – liquid solution is being prepared. For example, beer is about 5% ethanol by volume. This means every 100 mL beer contains 5 mL ethanol (ethyl alcohol).

### 1.3.3. Molarity (M)

It shows number of moles of a given substance per litre of solution. For instance: 4.0 litres of liquid, containing 2.0 moles of dissolved particles, constitutes a solution of 0.5 M. Such a solution may be described as "0.5 molar."

### 1.3.4. Molality (m)

It shows number of moles of a given substance per kilogram of solvent. For instance: 2.0 kilograms of solvent, containing 1.0 moles of dissolved particles, constitutes a molality of 0.5 mol/kg. Such a solution may be described as "0.5 molal." The advantage of molality is, it does not change with the temperature, as it deals with the mass of solvent rather than the volume of solution. Volume increases with increase in temperature resulting in decrease in molarity. Molality of a solution is always constant irrespective of the physical conditions like temperature and pressure.

### 1.3.5. Normality (N)

Normality is a concept related to molarity, usually applied to acid-base solutions and reactions. For acid-base reactions, the equivalent is the mass of acid or base that can accept or donate exactly one mole of protons (H+ ions). Normality is also used for redox reactions. In this case the equivalent is the quantity of oxidizing or reducing agent that can accept or furnish one mole of electrons. Whereas molarity measures the number of particles per litre of solution, normality measures the number of equivalents per litre of solution. In practice, this simply means one multiplies the molarity of a solution by the valence of the ionic solute. A bit more complex for redox reactions where the normality is always equal to, or greater than molarity for acid-base reactions. For redox reactions the normality is typically equal to or less than the molarity.

### 1.3.6. Formal (F)

The formal (F) is yet another measure of concentration similar to molarity. It is used rarely. It is calculated based on the formula weights of chemicals per litre of solution. The difference between formal and molar concentrations is that the formal concentration indicates moles of the original chemical formula in solution, without regard for the species that actually exist in solution.

Molar concentration, on the other hand, is the concentration of species in solution. For Example if one dissolves sodium carbonate ($Na_2CO_3$) in a litre of water, the compound dissociates into the $Na^+$ and $CO_3^{2-}$ ions. Some of the $CO_3^{2-}$ reacts with the water to form $HCO_3^-$ and $H_2CO_3$. If the pH of the solution is low, there is practically no $Na_2CO_3$ left in the solution. So, although we have added 1 mol of $Na_2CO_3$ to the solution, it does not contain 1 M of that substance. However, one can still say that the solution contains 1 F of $Na_2CO_3$.

# 2 Equilibrium Constant

## 2.1 Introduction

The equilibrium constant, K, expresses the relationship between products and reactants of a reaction at equilibrium with respect to a specific unit. There are two different types of equilibrium reactions:

1. Homogeneous
2. Heterogeneous

We would like to state the extent of a reaction which involves dynamic equilibrium in order to determine the factors that will affect the extent of the reaction. For this, the equilibrium constant, K is used which can be defined in terms of the equilibrium concentrations of the reactants and products for the reaction aA + bB → cC + dD:

$$K_c = \{[C]^c[D]^d / [A]^a[B]^b\}_{eq}$$

Here, c shows that the equilibrium constant can also be written in terms of partial pressures $K_p$, or in terms of mole fractions $K_x$. This constant is a function of temperature so, it can be written as K(T). If $K_c$ were equal to $6.6 \times 10^{40}$ as the value is very large, that means that the numerator is much larger than the denominator and that tells us that the reaction is strongly shifted toward the products. If $K_c$ were equal to $4.8 \times 10^{-25}$, then the denominator was much larger than the numerator and the reaction was shifted strongly toward the reactants.

Equilibrium constants $K_{eq}$, are found by multiplying the product concentrations raised to the power of their stoichiometric coefficients divided by the concentrations of reactants raised to the power of their stoichiometric coefficients. $K_{eq}$ measures the extent to which reactants are converted to products in a reaction. $K_{eq}$ is a number that, under specified conditions, is always the same for a given reaction, when the temperature changes, the equilibrium constant will change.

### 2.1.1 Homogeneous Reactions

The simpler reaction is homogeneous reaction which shows the states of matter of the products and reactions as similar. In most cases, the solvent determines the state of matter for the overall reaction. For example, the synthesis of methanol from a carbon monoxide-hydrogen mixture is a gaseous homogeneous mixture, which contains two or more substances:

$$CO(g) + 2H_{2(g)} \rightleftharpoons CH_3OH$$

At equilibrium, the rate of the forward and reverse reaction are equal, which is demonstrated by the arrows. The equilibrium constant, however, gives the ratio of the units (pressure or concentration) of the products to the reactants when the reaction is at equilibrium. The synthesis of ammonia is another example of a gaseous homogeneous mixture:

$$N_{2(g)}+3H_{2(g)}\rightleftharpoons 2NH_3$$

Consider a reaction involving a homogeneous system, such that their equilibrium is represented by the law of mass action:

$$wA + xB \rightleftharpoons yC + zD;\ K_c = ([C]^y[D]^z)/([A]^w[B]^x)$$

where,

A, B, C, and D are chemical species

w, x, y, and z are respective coefficients

The equilibrium constant, $K_c$, for similar systems, as shown in the examples below, can be expressed as follows:

a) $H_2(g) + I_2(g) \rightleftharpoons 2\,HI(g);\ K_c = \dfrac{[HI]^2}{[H_2][I_2]}$

b) $N_2(g) + 3H_2(g) \rightleftharpoons 2\,NH_3(g);\ \dfrac{[NH\ ]}{[N_2][H_2]}$

### 2.1.2 Heterogeneous Reaction

Heterogeneous reaction is that where one or more states within the reaction is different. For example, the formation of an aqueous solution of lead(II) iodide creates a heterogeneous mixture dealing with molecules in both the solid and aqueous states:

$$PbI_{2(s)} \rightleftharpoons Pb^{+2}(aq) + 2I^-(aq)$$

The decomposition of sodium hydrogen carbonate at high elevations is another example of a heterogeneous mixture, this reaction deals with molecules in both the solid and gaseous states:

$$2NaHCO_{3(s)} \rightleftharpoons Na_2CO_{3(s)} + H_2O_{(g)} + CO_{2(g)}$$

$$C_{(s)} + O_{2(g)} \rightleftharpoons CO_{2(g)}$$

Once again, this difference is emphasized so that students remember that equilibrium constant calculations are different from heterogeneous mixtures.

**Remember**

Note some of the special means of dealing with equilibrium constants:

- If a solid is involved as a product or a reactant, it does not appear in the equilibrium constant. So for the reaction $CaCO_3(s) \rightarrow CaO(s) + CO_2(g)$, the equilibrium constant would be $K_c = [CO_2(g)]$.
- For reactions involving pure liquid water in which the molar concentration of water is virtually unchanged during the course of the reaction, the water concentration does not appear in the equilibrium constant.
- If the equilibrium reaction is written in a reverse manner, the new equilibrium constant is the inverse of the original K. For the reaction aA + bB → cD + dD, the equilibrium constant is $K_c = [C]^c[D]^d / [A]^a[B]^b$
- If we reverse the reaction: cD + dD → aA + bB, then the equilibrium constant is inverted

  $K'_c = 1/K_c = [A]^a[B]^b/[C]^c[D]^d$
- If we multiply both sides of the reaction by any factor (2, 3, ½, 1/3 , etc) then the equilibrium constant is raised to that power. For example, if we multiply the reaction aA → bB through by 2, the new equilibrium constant will be the square of the original. If we multiply the same equation by ½, the new K is the square root of the original

### 2.1.3 Law of Mass Action

Heat is observed as flow of energy from high temperature object to low temperature object. In such case, when two objects are at same temperature, then there will be no net flow of energy or heat. With this, a covered cup of coffee will not be cooler or warmer than the room temperature after it has been in there for a few hours, such phenomenon is equilibrium. In this example, we deal with the flow of energy.

Equilibria happen in phase transitions. When the temperature in a system having a mixture of ice and water is uniformly 273.15 K, the net amount of ice formed and the melt will be zero. The amount of liquid water will also remain constant, when no vapour escape from the system. In this case, three phases, ice (solid) water (liquid), and vapour (gas) are in equilibrium with one another. Also, equilibrium also establishes between the vapour phase and liquid at a particular temperature. Equilibrium conditions also exist between solid phase and vapour phases. These are phase equilibria.

Chemical reactions may not be as complete as it is assumed in Stoichiometry calculations. For example, in the following reaction with short completion:

$$2NO_2 = N_2O_4$$

$$3H_2 + N_2 = 2\ NH_3$$

$$H_2O + CO = H_2 + CO_2$$

first reaction at room temperature, is not possible for pure $NO_2$ or $N_2O_4$ as it is in a sealed tube or closed system with ratio being constant.

$$[N_2O_4]/[NO_2]^2$$

Such phenomenon is known as chemical equilibrium and law of nature is known as law of mass action or mass action law. Further, when conditions like pressure and temperature changes, a period of time is required for the system to establish in equilibrium.

Before knowing mass action law, it is important to find a system or closed system. The law provides an expression for a constant for all reversible reactions. For systems which is not in equilibrium, ratio calculated from mass action will be reaction quotient Q. The Q values of a closed system have a tendency to reach a limiting value called equilibrium constant K over time. A system has a tendency to reach an equilibrium state.

The law of mass action is universal, applicable under any circumstance. However, for reactions that are complete, the result may not be very useful. We introduce the mass action law by using a general chemical reaction equation in which reactants A and B react to give product C and D.

$$a\,\mathrm{A} + b\,\mathrm{B} \rightarrow c\,\mathrm{C} + d\,\mathrm{D}$$

where *a, b, c, d* are the coefficients for a balanced chemical equation.

The mass action law states that if the system is at equilibrium at a given temperature, then the following ratio is a constant.

$$\frac{[\mathrm{C}]^c\,[\mathrm{D}]^d}{[\mathrm{A}]^a\,[\mathrm{B}]^b} = \mathrm{K}_{eq}$$

The square brackets "[ ]" around the chemical species represent their concentrations. This is the ideal law of chemical equilibrium or law of mass action. The units for *K* depend upon the units used for concentrations. If M is used for all concentrations, *K* has units

$$\mathrm{M}^{c+d-(a+b)}$$

If the system is NOT at equilibrium, the ratio is different from the equilibrium constant. In such cases, the ratio is called a reaction quotient which is designated as *Q*.

$$\frac{[\mathrm{C}]^c\,[\mathrm{D}]^d}{[\mathrm{A}]^a\,[\mathrm{B}]^b} = \mathrm{Q}$$

A system not at equilibrium tend to become equilibrium, and the changes will cause changes in *Q* that its value approaches the equilibrium constant, *K*

$Q \rightarrow K_{eq.}$

The mass action law gives us a general method to write the expression for the equilibrium constant of any reaction. At this stage, you should be able to write the equilibrium expression for any reaction equation. If you are not sure from the above general theory, here are some examples.

**Examples**

1. 1.The equilibrium constant expression for reaction equation $NH_3 + HOAc = NH_4^+ + OAc^-$ will be:

$$\frac{[NH_4^+][OAc^-]}{[NH_3][HOAc]} = K\left(\text{unitless constant}\right)$$

2. In case of ionization of acid, $H_2SO_4 = 2\ H^+ + SO_4^{2-}$ the equilibrium constant expression will be equilibrium constant as:

$$\frac{[H^+]^2[SO_4^{2-}]}{[H_2SO_4]} = K\ M^2$$

where M = mol/L. Note the unit for *K*.

3. In case of reaction equation $Cu^{2+} + 6\ NH_3 = Cu(NH_3)_6^{2+}$ the equilibrium constant expression is:

$$\frac{[Cu(nH\ )^{2+}]}{[Cu^{2+}][NH\ ]} \qquad M$$

The equilibrium constant depends on the written equation.

4. In the ionization of oxalic acid, $H_2C_2O_4$, it has been shown that $H_2C_2O_4 = H^+ + HC_2O_4^-$ where

$$\frac{[HC_2O_4^-][H^+]}{[H_2C_2O_4]} = 0.059\ M$$

The second ionization constant is much smaller so $HC_2O_4^- = H^+ + C_2O_4^{2-}$

$$\frac{[H^+][C_2O_4^{2-}]}{[HC_2O_4^-]} = K_2 = 0.000064\ M$$

The total ionization can be obtained from above two as $H_2C_2O_4 = 2\ H^+ + C_2O_4^{2-}$ and the equilibrium constant is

$$\frac{\left[H^{+}\right]^{2}\left[C_2O_4^{\,2-}\right]}{\left[H_2C_2O_4\right]}=K_{overall}\ M^{2}$$

Further $K_{overall} = K_1 * K_2$

$= 3.8x10^{-6} M^2$

5. The equilibrium constant expression and value for reaction $CH_4(g)$ + $H_2O(g) = CO(g) + 3\ H_2(g)$ will be shown as:

$$\frac{[CO][H_2]^3}{[CH_4][H_2O]} = \frac{1}{4.0} = 0.25$$

6. At some temperature, equilibrium constant is 4.0 for reaction $CO(g) + 3\ H_2(g) = CH_4(g) + H_2O(g)$ while the equilibrium constant expression and value for reaction $1/3\ CO(g) + H_2(g) = 1/3\ CH_4(g) + 1/3\ H_2O(g)$ result as:

$$\frac{[CH_4]^{1/3}[H_2O]^{1/3}}{[CO]^{1/3}[H_2]} = 4.0^{1/3} = 1.59$$

The application of the mass action law leads to the method to write the expression for the equilibrium constant. The law is given in a general form, and these examples should help you grasp the method.

## 2.2 Determining Equilibrium Constant, $K_c$

During reaction of chemical substances, reaction normally does not go to completion, but instead, the system reached to certain intermediate state where rates of the forward and reverse reactions are equal.

In such state, reactants and products have concentrations that do not change with time. This type of system is said to be in chemical equilibrium where at particular temperature, reaction mixture obeys Law of Chemical Equilibrium, which imposes a condition on the concentrations of reactants and products. This condition is expressed in the equilibrium constant $K_C$ for the reaction. In an experiment, study of equilibrium properties of reaction between iron(III) ion and thiocyanate ion are considered:

$$Fe^{3+}\ (aq) + SCN^{-}\ (aq) \rightleftharpoons FeSCN^{2+}\ (aq)$$

When solutions having $Fe^{3+}$ ion and thiocyanate ion are mixed, they react forming $FeSCN^{2+}$ complex ion, having deep red color. With this reaction, equilibrium amounts of $Fe^{3+}$ and $SCN^-$ will be less as they would have been when no reaction occurs or takes place, for every mole of $FeSCN^{2+}$ formation having 1mole of $Fe^{3+}$ and 1mole of $SCN^-$. The equilibrium constant expression $K_C$ for Reaction 1 is:

$$\frac{\left[FeSCN^{2+}\right]}{\left[Fe^{3+}\right]\left[SCN^{-}\right]} = K_c$$

Here, value of $K_C$ is constant at given temperature that shows that mixtures containing $Fe^{3+}$ and $SCN^-$ will result in equilibrium with same value of $K_C$, no matter what initial amounts of $Fe^{3+}$ and $SCN^-$ were used. The idea of this experiment is to find $K_C$ for the reaction for several mixtures that have been made up in different ways, and to show that $K_C$ indeed has the same value in each of the mixtures. The reaction is a particularly good one to study because $K_C$ is of a convenient magnitude and the color of $FeSCN^{2+}$ ion makes for an easy analysis of the equilibrium mixture.

The mixtures will be prepared by mixing solutions containing known concentrations of iron (III) nitrate, $Fe(NO_3)_3$, and potassium thiocyanate, KSCN. The color of the $FeSCN^{2+}$ ion formed will allow us to determine its equilibrium concentration. Knowing the initial composition of a mixture and the equilibrium concentration of $FeSCN^{2+}$, we can calculate the equilibrium concentrations of the rest of the pertinent species and then determine $K_c$.

**Determining [$FeSCN^{2+}$]**

In an experiment, a spectrophotometer is applied to find [$FeSCN^{2+}$] in equilibrium mixtures where instructions for use of spectophotometer are detailed. The spectrophotometer measures absorbance, which is the amount of light absorbed by complex at given wavelength. As per Beer's law, relationship between absorbance A and concentration of colored species c will result as

$$A = \varepsilon bc$$

Here,

$\varepsilon$ = constant that depends on wavelength of light and on substance which absorbs light

b = distance that light travels through sample of absorbing substance

The $FeSCN^{2+}$ complex absorbs blue light thereby giving reddish orange color. It absorbs the most light at a wavelength of 447 nm. Therefore, at this wavelength absorbance measurements will have the highest sensitivity to [$FeSCN^{2+}$]. In this experiment, you will measure the absorbance of all solutions at 447 nm.

During the first part of the experiment, it is required to get the relation among absorbance and [$FeSCN^{2+}$] at 447 nm. For this, measuring of absorbance of three standard solutions is required, in which [$FeSCN^{2+}$] is known. It is noticed that a problem occurs as how can a known concentration of $FeSCN^{2+}$ be obtained as $FeSCN^{2+}$ participates in equilibrium with $Fe^{3+}$ and $SCN^-$ ions. The known amounts of reactants will not necessarily give required amount of product.

As per Le Châtelier's principle, a net reaction from left to right (that is, in the forward direction) occurs when more of a reactant is added. As more and more of the same reactant is added, more and more of the product will be formed. It is possible to add so much of this reactant that essentially all of the other reactant will be converted to product. By considering the limiting quantities of $SCN^-$ and overwhelming amounts of $Fe^{3+}$ to get known amounts of $FeSCN^{2+}$ in standard solutions, the amount of $FeSCN^{2+}$ which is formed will be essentially same as starting amount of limiting reactant.

For this, measure the absorbance's of standard solutions by plotting the absorbance against the concentrations of $FeSCN^{2+}$ on graph or calibration curve. The points on calibration curve will fall on straight line having a slope $\varepsilon b$, where constant of proportionality among absorbance and concentration is seen.

In the second part of experiment, absorbance of different set of solutions is measured where substantial amounts of reactants and product are available. Here the calibration curve is applied to convert measured absorbance to equilibrium concentration of $FeSCN^{2+}$ in every solution. From the initial concentrations of the reactants and the equilibrium concentration of the product, you will calculate the equilibrium constant for the reaction.

In preparing each of the mixtures in this experiment you will maintain the concentration of $H^+$ ion at 0.5 M. The hydrogen ion does not participate directly in the reaction, but its presence is necessary to avoid the formation of brown-colored species such as $FeOH^{2+}$, which would interfere with the analysis of $[FeSCN^{2+}]$.

**Example**

Suppose that 0.1000 mole each of $H_2$ and $I_2$ are placed in 1.000-L flask, stoppered, and the mixture is heated to 425°C. At equilibrium, the concentration of $I_2$ is found to be 0.0210 M. What are the equilibrium concentrations of $H_2$ and HI, respectively? Calculate $K_c$ for the following reaction at 425°C.

$$H_2(g) + I_2(g) \rightleftharpoons 2\,HI(g)$$

Calculate the concentration of $H_2$, $I_2$ and HI at equilibrium using the ICE table:

| | $H_2(g)$ | + $I_2(g)$ | $\rightleftharpoons$ 2 HI(g) |
|---|---|---|---|
| Initial [ ], M: | 0.1000 | 0.1000 | 0.0000 |
| Change in [ ], M: | -0.0790 | -0.0790 | +(2 × 0.0790) |
| Equilibrium [ ], M | 0.0210 | 0.0210 | 0.1580 |

$$K_c = \frac{[HI]^2}{[H_2][I_2]} = \frac{(0.1580)^2}{(0.0210)^2} = 56.6$$

For a given reaction at constant temperature, the value of $K_c$ does not depend on the amounts of reactants that are initially mixed, nor on any specific set of concentrations at equilibrium. For a given reaction, different sets of concentrations may occur at equilibrium, but the same value of $K_c$ will be obtained at a given temperature. $K_c$ depends on how the chemical equation is written, but not on how the reaction is carried out.

For example, supposed that 0.05000 mole of HI is placed in a 1.000-L flask, and the mixture is sealed and heated to 425°C. At equilibrium, the concentration of $I_2$ is found to be 0.00525 M. Calculate $[H_2]$ and [HI] at equilibrium, and the value of $K_c$ for the reaction:

$$H_2(g) + I_2(g) \rightleftharpoons 2\,HI(g)$$

Calculate the concentration of $H_2$, $I_2$ and HI at equilibrium using the **ICE** table:

| | $H_2(g)$ | + | $I_2(g)$ | $\rightleftharpoons$ | 2 HI(g) |
|---|---|---|---|---|---|
| Initial [ ], M: | 0.00000 | | 0.00000 | | 0.05000 |
| Change in [ ], M: | +0.00525 | +0.00525 | -(2 × 0.00525) | | |
| Equilibrium [ ], M | 0.00525 | 0.00525 | 0.03950 | | |

$$K_c = \frac{(0.0395)^2}{(0.00525)^2} = 56.6$$

The initial reaction may occur in the reverse direction and the equilibrium mixture may contain a different set of concentrations, but the value of $K_c$ for the reaction at any given temperature will remain constant. Each set of equilibrium concentrations is called equilibrium position. Thus, a given reaction at a fixed temperature has only one $K_c$ value, but it can have a finite number of equilibrium positions.

## 2.2 Forms of Equilibrium Constants

### 2.2.1 Equilibrium Constant of Concentration

The equilibrium constant of concentration gives the ratio of concentrations of products over reactants for a reaction that is at equilibrium. This is usually used when the state of matter for the reaction is (aq). The equilibrium constant expression is written as Kc, as in the expression below:

Kc=[G]g[H]h[A]a[B]b

If *K*>1K>1 then equilibrium favors products

If *K*<1K<1 then equilibrium favors the reactants

Here, the letters inside the brackets represent the concentration of each molecule. It is seen that the mathematical product of chemical products raised to powers of respective coefficients is the numerator of the ratio and the mathematical product of the reactants raised to the powers of their respective coefficients is the denominator. This is the case for every equilibrium constant. Keep in mind that this expression was obtained by a homogeneous equilibrium reaction. K represents an equilibrium constant and *c*c represents concentration (e.g., Kc). This means that every species shows up in the expression, as long as it is a solution or a gas.

**Example**

Given the reaction A ⇔ B + C, find the equilibrium concentrations of B and C if initial concentration of A is 1.00 M and equilibrium constant is $1.86 \times 10^{-6}$?

**Solution**

Consider the equation and writing the first step of equilibrium expression for the process:

$$K_{eq} = {}^{[B][C]}/_{[A]}$$

Now, finding the concentration of each species at equilibrium by setting up a chart that shows the initial concentrations of species by changing the concentrations of each for having final concentrations of each species. For this, consider the chart:

| Species | Initial Concentration | Change | Final Concentration |
|---|---|---|---|
| A | 1.00 M | -x | (1.00 - x) M |
| B | 0 M | x | x M |
| C | 0 M | x | x M |

Here, initial concentration of A is defined by problem as 1.00 M. Between the time that the reaction starts and the system reaches equilibrium, some of it will turn into the products. As a result, our final concentration will be the initial concentration minus the amount of change, or (1.00 - x) M.

The initial concentrations of B and C are zero as neither was initially present, but from the equation, every time a molecule of A breaks apart forming one molecule each of B and C. As a result, if concentration of A decreases by x, concentration of products increases by x. At equilibrium, concentration of both B and C will be x.

For locating the concentrations of each species, put these values in expression for finding the equilibrium constant. As it is known that equilibrium constant is $1.86 \times 10^{-6}$, so solving for x in the expression:

$$1.86 \times 10^{-6} = {}^{[x][x]}/_{[1.00 - x]}$$

Considering the quadratic equation when $K_{eq}$ values are very small, when $K_{eq}$ is small, then we guess that the amount of product formed x will be very small when compared to initial quantity of reactant. It is because a very small $K_{eq}$ value means very little product has been formed. As a result, we can omit the "x" in the denominator to simplify the [1.00 - x] term because [1.00 - x] will be roughly equal to 1.00.

$$1.86\times10^{-6} = {[x][x]}/_{[1.00]}$$

$$x = 1.36 \times 10^{-3}\ M$$

If the amount of product obtained as $1.36 \times 10^{-3}$ M, then final concentration of acetic acid will be (1.00 - 0.00136) M which is within the constraints of significant figure rules that is 1.00 M. Hence the final concentrations of B and C will be $1.36 \times 10^{-3}$ M.

## 2.2.2 Equilibrium Constant of Pressure

When the reactants and products are in gaseous state then the concentrations are expressed in terms of partial pressure instead of the molar concentrations then the equilibrium constant is termed as equilibrium constant in terms of partial pressure and is represented by $K_p$

For gas-specific reactions, however, we can express the equilibrium constant in terms of the partial pressures of the gases involved. Take the general gas-phase reaction:

$$aA(g)+bB(g) \rightleftharpoons cC(g)+dD(g)$$

Our equilibrium constant in terms of partial pressures, designated $K_P$, is given as:

$$KP=P^{C}_{C}\,P^{d}_{D} / P^{a}_{A}\,P^{b}_{B}$$

Note that this expression is extremely similar to $K_C$, the equilibrium expression written in terms of concentrations. In order to prevent confusion, do not use brackets ([ ]), when writing $K_P$ expressions.

**Example**

When 0.40 moles of $PCl_5$ is heated in a 10.0 L container, an equilibrium is established is which 0.25 moles of $Cl_2$ is present. $PCl_5(g) \Leftrightarrow PCl_3(g) + Cl_2(g)$. Find the number of moles of $PCl_5$ and $PCl_3$ **at equilibrium** and equilibrium concentration of all three components?

| $PCl_5$ | $\Leftrightarrow$ | $PCl_3$ | + | $Cl_2$ |
|---|---|---|---|---|
| start 0.40 moles | | ------- | | ------ |
| after 0.15 moles | | 0.25 moles | | 0.25 moles |

**Solution**

From the balanced equation and the coefficients it should be obvious that everything works on a one-to-one basis. So there should be 0.25 moles of $PCl_3$ **and 0.25 moles of $PCl_5$ should have been consumed, so there are 0.40-0.25 moles = 0.15** moles of $PCl_5$ left.

$$\text{Therefore } [PCl_5] = \frac{0.15 \text{ moles}}{10.0 \text{ L}} = 0.015 \text{ moles/L}$$

$[PCl_3]$ and $[Cl_2]$ both are 0.25 moles each. Therefore their concentrations are identical as well.

$$[PCl_3] = [Cl_2] = \frac{0.25 \text{ moles}}{10.0 \text{ L}} = 0.025 \text{ moles/L}$$

**Example**

When 3.0 moles of HI, 20 moles of $H_2$, and 1.5 moles of $I_2$ **are placed in a 1.0 L container at 448**°C, will a reaction occur? Calculate the concentrations and substitute them into an expression and compute the "experimental concentration quotient" or "Q".

**Solution**

If Q is equal to $K_{eq}$ **(50 in this case) then the system is already at equilibrium and stable.**

If $Q > K_{eq}$ **then the [product] is too high and must decrease. Therefore the reverse reaction is favoured.**

If $Q < K_{eq}$ **then the [product] is still to small and must increase, therefore the forward reaction is favoured.**

$$H_2 + I_2 \Leftrightarrow 2\,HI$$

$$Keq = [HI]^2/[I_2]\,[H_2]$$

$$Q = \frac{(3.0 \text{ mol/L})^2}{(2.0 \text{ mol/L})(1.5 \text{ mol/L})}$$

$$= \frac{9.0}{3.0} = 3$$

Hence, $Q < K_{eq}$ as $3 < 50$ and forward reaction exists till the ratio of product: reactants is 50.

### 2.2.3 Relationship between $K_p$ and $K_c$

Kc and Kp are the equilibrium constants of gaseous mixtures. However, the difference between the two constants is that Kc is defined by molar concentrations, whereas Kp is defined by the partial pressures of the gasses inside a closed system. The equilibrium constants do not include the concentrations of single components such as liquids and solid, and they do not have any units.

$$Kp = Kc\,(RT)\,\Delta n$$

Kp and Kc are two equilibrium constants for a reversible reaction, aA + bB → cC + dD

According to the Law of active masses by Goldberg and Waage,

Kc=products of concentration of products/ products of concentration of reactants

Kc=[C]^c•[C]^d / [C]^a•[C]^b

Where

[C] = concentration of reactants and products.

Kp=product of the partial pressure of products/product of the partial pressure of reactants

Kp=[P]^c•[P]^d / [P]^a•[P]^b

The relation between Kp and Kc is derived by the substitution of the formula for pressure (P),

PV=nRT

P=(n/V)RT

P=CRT

Therefore, the relation between Kp and Kc can be derived by this formula viz.

Kp = Kc • (RT)^Δng

Where, R is constant (0.0821 bar•litre/mol•K), T is temperature and Δng is the difference of moles, Δng=moles of products - moles of reactants viz. Δng=(c+d)-(a+b).

Consider the equilibrium 2 $SO_2(g) + O_2(g) \rightleftharpoons 2\ SO_3(g)$;

$$K_c = \frac{[SO]^2}{[SO_2]^2[O_2]}$$

$$K_p = \frac{(P_{SO3})^2}{(P_{SO2})^2(P_{O2})}$$

Assuming ideal gas behavior: PV = nRT, we derive the following relationships:

$P_{SO_3} = (n_{SO_3})RT/V$;

$P_{SO_2} = (n_{SO_2})RT/V$;

$P_{O_2} = (n_{O_2})RT/V$

$(n_{SO_3})/V = (P_{SO_3})/RT = [SO_3]$;

$(P_{SO_2})/RT = [SO_2]$;

$(P_{O_2})/RT = [O_2]$

$$K_c = \frac{[SO]^2}{[SO_2]^2[O_2]}$$

$$= \frac{(P_{SO3}/RT)^2}{(P_{SO2}/RT)^2(P_{O2}/RT)}$$

$$= \frac{(P_{SO3})^2}{(P_{SO2})^2(P_{O2})} \times (RT)$$

$$= K_p(RT)$$

For the reaction: $2\ SO_2(g) + O_2(g) \rightleftharpoons 2\ SO_3(g)$, $K_c = K_p(RT)$, or $K_p = K_c(RT)^{-1}$

In general, for reactions involving gases, such as:

$wA(g) + xB(g) \rightleftharpoons yC(g) + zD(g)$; $\mathbf{K_p = K_c(RT)^{\Delta n}}$;

where $\Delta n = (y + z) - (w + x)$; (w, x, y, and z are coefficients of gas species)

**Examples**

1. $PCl_5(g) \rightleftharpoons PCl_3(g) + Cl_2(g)$; $K_p = K_c(RT)$;
2. $2NO_2(g) \rightleftharpoons N_2O_4(g)$; $K_p = K_c(RT)^{-1}$
3. $H_2(g) + I_2(g) \rightleftharpoons 2\ HI(g)$; $K_p = K_c$;
4. $N_2(g) + 3H_2(g) \rightleftharpoons 2\ NH_3(g)$; $K_p = K_c(RT)^{-2}$

## 2.3 Dissociation Constants

Dissociation constant $K_d$ is particular type of equilibrium constant that measures the propensity of a larger object to separate (dissociate) reversibly into smaller components, as when a complex falls apart into its component molecules, or when a salt splits up into its component ions. The dissociation constant is the inverse of the association constant. In the special case of salts, the dissociation constant can also be called an ionization constant.

The dissociation constant specifies the tendency of a substance $A_x B_y$ to reversibly dissociate (separate) in a solution into smaller components A and B:

$AyBy \rightleftharpoons xA+yB$

The dissociation constant is denoted $K_d$ and is calculated by

$Kd=[A]^x [B]^y / [A_x B_y]$

where

[A], [B], and $[A_xB_y]$ are the molar concentrations of entities A, B, and $A_xB_y$.

The dissociation constant is an immediate consequence of the law of mass action which describes equilibria in a more general way. The dissociation constant is also sometimes called ionization constant when applied to salts. The inverse of the dissociation constant is called association constant.

### 2.3.1 Dissociation Constants of Acids and Bases

#### Dissociation Constant for Weak Acid

Consider an acid *HA* which, when dissolved in water ionizes as,

$$HA \rightleftharpoons H^+ + A^-$$

Applying the law of mass action,

$$K_a = \frac{[H^+][A^-]}{[HA]}$$

Where,

$K_a$ = dissociation constant of acid *HA*

It has constant value at definite temperature and does not change with the change of concentration.

#### Dissociation Constant for Polybasic Acid

Polybasic acids ionise stepwise as, for example, orthophosphoric acid ionises in three steps and each step has its own ionisation constant.

$$H_1PO_4 \rightleftharpoons H^+ + H_1PO_4 \text{(I step)}$$

$$H_1PO_4^- \rightleftharpoons H^+ + HPO_4^{-1} \text{(II step)}$$

$$HPO_4^{-2} \rightleftharpoons H^+ + PO_4^{-1} \rightleftharpoons \text{(III step)}$$

Let $K_1, K_2$ and $K_1$ be the ionization constants of first, second and third steps respectively. Thus,

$$K_1 = \frac{[H^+][H_1PO_4^-]}{[H_2PO_4]}$$

$$K_1 = \frac{[H^+][HPO_4^{-2}]}{[H_2PO_4^-]}$$

$$K_1 = \frac{[H^+][PO_4^{-1}]}{[HPO_4^{-2}]}$$

In general, $K_1 > K_2 > K_1$

The overall dissociation constant *(K)* is given by the relation,

$$K = K_1 \times K_2 \times K_1$$

Dissociation constant for weak base:

The equilibrium of $NH_4OH$ (a weak base) can be represented as,

$$NH_4OH \rightleftharpoons NH_4^+ + OH^-$$

Applying the law of mass action,

$$K_b = \frac{[NH_4^+][OH^-]}{[NH_4OH]}$$

$K_{b\,=}$ constant at a definite temperature and does not change with the change of concentration.

### 2.3.2 Dissociation Constant of Water

Formally, the dissociation of water follows the following equation:

$$H_2O \rightleftharpoons H^+ + OH^-$$

Thus the dissociation constant is given by

$$K_d = [H+][OH-]/[H_2O]$$

$$=2.16\times10^{-16}$$

However, since the concentration of undissociated water is almost unchanged by the dissociation process (due to the low $K_d$), we can assume that a liter of water at 25°C contains 55.39 mol undissociated water. For the sake of convenience this constant water concentration is combined with the dissociation constant $K_d$ to form the dissociation constant of water $K_w$:

$$K_w=[H^+][OH^-]$$

$$= 2.16\times10^{-16} \times 55.39$$

$$= 1.2\times10^{-14}$$

The value of $K_w$ changes considerably with temperature. Consequently this variation must be taken into account when making precise measurements (i.e. when determining the pH).

| Water Temperature [°C] | $K_w$ [$10^{-14}$] | $pK_w$ |
|---|---|---|
| 0 | 0.1 | 14.92 |
| 10 | 0.3 | 14.52 |
| 18 | 0.7 | 14.16 |
| 25 | 1.2 | 13.92 |
| 30 | 1.8 | 13.75 |
| 50 | 8.0 | 13.10 |
| 60 | 12.6 | 12.90 |
| 70 | 21.2 | 12.67 |
| 80 | 35 | 12.46 |
| 90 | 53 | 12.28 |
| 100 | 73 | 12.14 |

**Example**

A 0.031 mol/L solution of an unknown monoprotic acid has a pH of 4.69. What is the acid dissociation constant for this acid?

**Solution**

Writing equation for dissociation of generic monoprotic acid as:

$$HA + H_2O \rightleftharpoons H_3O^+ + A^-$$

Writing dissociation constant expression:

$$Ka=[H_3O^+][A^-]/[HA]$$

Determining equilibrium concentrations:

$$pH = -\log[H_3O^+] = 4.69$$

$$[H_3O^+] = 10-4.69 \text{ mol/L} = 2.04 \times 10^{-5} \text{ mol/L}$$

From the equation,

$$[A^-] = [H_3O^+] = 2.04 \times 10^{-5} \text{ mol/L}$$

$$[HA] = (0.031 - 2.04 \times 10^{-5}) \text{ mol/L}$$

$$= 0.031 \text{ mol/L}$$

Substitute these values in expression

$Ka=[H_3O^+][A^-][HA]$

$=2.04\times10^{-5}\times2.04\times10^{-5}0.031$

$= 1.34 \times 10^{-8}$

**Example**

A 0.150 M solution of acetic acid (HAc) is found to be 1.086% dissociated. What is $K_a$?

**Solution**

The equation for dissociation of HAc is:

$HAc \rightleftharpoons H^+ + Ac^-$

The $K_a$ expression is:

$K_a = ([H^+] [Ac^-]) / [HAc]$

Knowing about three values on right:

Finding $[H^+]$ using concentration and percent dissociation:

(0.01086) (0.15)

$= 1.629 \times 10^{-3}$ M

For $[Ac^-]$, using 1:1 stoichiometry from above equation:

$[Ac^-] = [H^+]$

$= 1.629 \times 10^{-3}$ M

For [HAc], using concentration given by subtracting $1.629 \times 10^{-3}$ from it.

Inserting in $K_a$ expression and solving further:

$x = [(1.629 \times 10^{-3}) (1.629 \times 10^{-3})] / 0.15$

$x = 1.77 \times 10^{-5}$

**Example**

A 1.3 M solution of the weak base methylamine ($CH_3NH_2$) has a percent ionization of 0.72%. What is its $K_b$?

**Solution**

Now writing ionization equation for methylamine and its $K_b$ expression:

$CH_3NH_2 + H_2O \rightleftharpoons CH_3NH_3^+ + OH^-$

Use B in place of $CH_3NH_2$:

$$K_b = ([HB^+]\,[OH^-]) / [B]$$

Using percent ionization to find $[OH^-]$ by:

$$(1.3)\,(0.0072) = 0.00936\ M$$

This is also $[HB^+]$

Calculating the value for $K_b$:

$$K_b = [(0.00936)\,(0.00936)] / 1.3$$

$$K_b = 6.7 \times 10^{-5}$$

## 2.4. Formation Constants of Complex Ion

Formation constant is equilibrium constant of reaction where complex ion are formed. On considering an aqueous solution of metal ion $M^{n+}$, it is observed that it will contain complex ion $[M(OH_2)_6]^{n+}$, so a sequence of substitution reactions will takes place on adding another ligand L. For all six steps, it is required to find equilibrium constant K.

Now on abbreviating molecules by leaving the water ligands, charges and square brackets, so that M means $[M(OH_2)_6]^{n+}$, ML means $[ML(OH_2)_5]^{n+}$, and so on. In an expressions for equilibrium constants, square brackets have their own meaning of concentration of particular species. As water is solvent and the reactants, its concentration will be taken ⇌ constant, hence considering following expressions.

$$M + L \square\ ML \qquad \rightleftharpoons$$

$$ML + L \square\ ML_2 \qquad \rightleftharpoons$$

$$ML_2 + L \square\ ML_3 \qquad K_3 = \frac{[ML_3]}{[ML_2][L]}$$

$$\rightleftharpoons$$

$$ML_N + L \square\ ML_N \qquad K_N = \frac{[ML_N]}{[ML_{N-1}][L]}$$

In the above expressions, $K_1$ to $K_N$ are stepwise stability constants or stepwise formation constants, so considering the reaction:

$$M + L \rightleftharpoons ML \qquad \beta_1 = \frac{[ML]}{[M][L]}$$

$$M + 2L \rightleftharpoons ML_2 \qquad \beta_2 = \frac{[ML_2]}{[M][L]^2}$$

$$M + 3L \rightleftharpoons ML_3 \qquad \beta_2 = \frac{[ML_2]}{[M][L]^3}$$

$$M + NL \rightleftharpoons ML_N \qquad 2 \quad \frac{[ML\ \ ]}{[M][L]}$$

Here, $\beta_1$ to $\beta_N$ are overall stability constants which are applied for relationship among two types of stability constants:

$$\beta_1 = K_1$$
$$\beta_2 = K_1 K_2$$
$$\beta_3 = K_1 K_2 K_3$$
$$\beta_N = K_1 K_2 K_3 \ldots K_N$$
$$\beta_2 = K_1 K_2 = \frac{[ML][ML_2]}{[M][L][ML][L]} = \frac{[ML_2]}{[M][L]^2}$$

| Complex Ion Equilibrium | $K_{form}$ |
|---|---|
| Halide complexes | |
| $Al^{3+} + 6\,F^- \leftrightarrow [AlF_6]^{3-}$ | $2.5 \times 10^4$ |
| $Al^{3+} + 4\,F^- \leftrightarrow [AlF_4]^{-1}$ | $2.0 \times 10^8$ |
| $Be^{2+} + 4\,F^- \leftrightarrow [BeF_4]^{2-}$ | $1.3 \times 10^{13}$ |
| $Sn^{4+} + 6\,F^-$ v $\leftrightarrow [SnF_6]^{2-}$ | $1.0 \times 10^{25}$ |
| $Cu^+ + 2\,Cl^- \leftrightarrow [CuCl_2]^{-1}$ | $3.0 \times 10^5$ |
| $Ag^+ + 2\,Cl^- \leftrightarrow [AgCl_2]^{-1}$ | $1.8 \times 10^5$ |
| $Pb^{2+} + 4\,Cl^- \leftrightarrow [PbCl_4]^{2-}$ | $2.5 \times 10^{15}$ |
| $Zn^{2+} + 4\,Cl^- \leftrightarrow [ZnCl_4]^{2-}$ | 1.6 |
| $Hg^{2+} + 4\,Cl^- \leftrightarrow [HgCl_4]^{2-}$ | $5.0 \times 10^{15}$ |
| $Cu^+ + 2\,Br^- \leftrightarrow [CuBr_2]^{-1}$ | $8.0 \times 10^5$ |
| $Ag^+ + 2\,Br^- \leftrightarrow [AgBr_2]^{-1}$ | $1.0 \times 10^{11}$ |
| $Hg^{2+} + 4\,Br^- \leftrightarrow [HgBr_4]^{2-}$ | $3.0 \times 10^4$ |
| $Cu^+ + 2\,I^- \leftrightarrow [CuI_2]^{-1}$ | $8.0 \times 10^8$ |
| $Ag^+ + 2\,I^- \leftrightarrow [AgI_2]^{-1}$ | $1.0 \times 10^{11}$ |
| $Pb^{2+} + 4\,I^- \leftrightarrow [PbI_4]^{2-}$ | $3.0 \times 10^4$ |
| $Hg^{2+} + 4\,I^- \leftrightarrow [HgI_4]^{2-}$ | $1.9 \times 10^{30}$ |

| | |
|---|---|
| Ammonia complexes | |
| $Ag^+ + \mathbf{2\ NH_3} \leftrightarrow \mathbf{[Ag(NH_3)_2]^+}$ | $1.6 \times 10^7$ |
| $Zn^{2+} + \mathbf{4\ NH_3} \leftrightarrow \mathbf{[Zn(NH_3)_4]^{2+}}$ | $7.8 \times 10^8$ |
| $Cu^{2+} + \mathbf{4\ NH_3} \leftrightarrow \mathbf{[Cu(NH_3)_4]^{2+}}$ | $1.1 \times 10^{13}$ |
| $Hg^{2+} + \mathbf{4\ NH_3} \leftrightarrow \mathbf{[Hg(NH_3)_4]^{2+}}$ | $1.8 \times 10^{19}$ |
| $Co^{2+} + \mathbf{6\ NH_3} \leftrightarrow \mathbf{[Co(NH_3)_6]^{2+}}$ | $5.0 \times 10^4$ |
| $Co^{3+} + \mathbf{6\ NH_3} \leftrightarrow \mathbf{[Co(NH_3)_6]^{3+}}$ | $4.6 \times 10^{33}$ |
| $Cd^{2+} + \mathbf{6\ NH_3} \leftrightarrow \mathbf{[Cd(NH_3)_6]^{2+}}$ | $2.6 \times 10^5$ |
| $Ni^{2+} + \mathbf{6\ NH_3} \leftrightarrow \mathbf{[Ni(NH_3)_6]^{2+}}$ | $2.0 \times 10^8$ |
| Cyanide complexes | |
| $Fe^{2+} + \mathbf{6\ CN^-} \leftrightarrow \mathbf{[Fe(CN)_6]^{4-}}$ | $1.0 \times 10^{24}$ |
| $Fe^{3+} + \mathbf{6\ CN^-} \leftrightarrow \mathbf{[Fe(CN)_6]^{3-}}$ | $1.0 \times 10^{31}$ |
| $Ag^+ + \mathbf{2\ CN^-} \leftrightarrow \mathbf{[Ag(CN)_2]^{-1}}$ | $5.3 \times 10^{18}$ |
| $Cu^+ + \mathbf{2\ CN^-} \leftrightarrow \mathbf{[Cu(CN)_2]^{-1}}$ | $1.0 \times 10^{16}$ |
| $Cd^{2+} + \mathbf{4\ CN^-} \leftrightarrow \mathbf{[Cd(CN)_4]^{2-}}$ | $7.7 \times 10^{16}$ |
| $Au^+ + \mathbf{2\ CN^-} \leftrightarrow \mathbf{[Au(CN)_2]^{-1}}$ | $2.0 \times 10^{38}$ |
| Complexes with other monodentate ligands | |
| $Ag^+ + \mathbf{2\ CH_3}NH_2 \leftrightarrow \mathbf{[Ag(CH_3}NH_2)_2]^{+1}$ | $7.8 \times 10^6$ |
| $Cd^{2+} + \mathbf{4\ SCN^-} \leftrightarrow \mathbf{[Cd(SCN)_4]^{2-}}$ | $1.0 \times 10^3$ |
| $Cu^{2+}\ \mathbf{2\ SCN^-} \leftrightarrow \mathbf{[Cu(SCN)_2]}$ | $5.6 \times 10^3$ |
| $Fe^{3+}\ \mathbf{3\ SCN^-} \leftrightarrow \mathbf{[Fe(SCN)_3]}$ | $2.0 \times 10^6$ |
| $Hg^{2+}\ \mathbf{4\ SCN^-} \leftrightarrow \mathbf{[Hg(SCN)_4]^{2-}}$ | $5.0 \times 10^{21}$ |
| $Cu^{2+}\ \mathbf{4\ OH^-} \leftrightarrow \mathbf{[Cu(OH)_4]^{2-}}$ | $1.3 \times 10^{16}$ |
| $Zn^{2+}\ \mathbf{4\ OH^-} \leftrightarrow \mathbf{[Zn(OH)_4]^{2-}}$ | $2.0 \times 10^{20}$ |
| Complexes with bidentate ligands (en = ethylenediamine ) | |
| $Mn^{2+} + \mathbf{3\ en} \leftrightarrow \mathbf{[Mn(en)_3]^{2+}}$ | $6.5 \times 10^5$ |
| $Fe^{2+} + \mathbf{3\ en} \leftrightarrow \mathbf{[Fe(en)_3]^{2+}}$ | $5.2 \times 10^9$ |
| $Co^{2+} + \mathbf{3\ en} \leftrightarrow \mathbf{[Co(en)_3]^{2+}}$ | $1.3 \times 10^{14}$ |
| $Co^{3+} + \mathbf{3\ en} \leftrightarrow \mathbf{[Co(en)_3]^{3+}}$ | $4.8 \times 10^{48}$ |
| $Ni^{2+} + \mathbf{3\ en} \leftrightarrow \mathbf{[Ni(en)_3]^{2+}}$ | $4.1 \times 10^{17}$ |
| $Cu^{2+} + \mathbf{2\ en} \leftrightarrow \mathbf{[Cu(en)_2]^{2+}}$ | $3.5 \times 10^{19}$ |
| $Co^{2+} + \mathbf{3\ C_2O_4^{2-}} \leftrightarrow \mathbf{[Co(C_2O_4)_3]^{4-}}$ | $4.5 \times 10^6$ |
| $Fe^{3+} + \mathbf{3\ C_2O_4^{2-}} \leftrightarrow \mathbf{[Fe(C_2O_4)_3]^{3-}}$ | $3.3 \times 10^{20}$ |

# 3 Ionic Strength and Activity Coefficients

## 3.1 Activity vs. Concentration

Activity is the state or quality of being active; nimbleness; agility; vigorous action or operation; energy; active force; as, an increasing variety of human activities while concentration is the act, process or ability of concentrating; the process of becoming concentrated, or the state of being concentrated.

In general, activity is what is important when doing any equilibrium calculations. In all possible places - dissociation constants, solubility products, complexation constants, Nernst equation and so on - it is activity that is important, not the concentration.

Concentration is the physical amount present in the solution. It is usually measured in grams or moles per unit volume. It is easily calculated. Activity is the deviation from the theoretical ideal when determining how the actual concentration reacts in the solution. It is determined by experiments and is based upon the calculations of Activity Coefficients. The Activity is the Activity Coefficient times the actual Concentration of the reactant in solution.

The best example is the concentration of Hydrogen ions in very strong acid solutions. The Activity (apparent concentration) of the Hydrogen ions (as measured by reactions or a pH meter) is usually less than the actual Concentration of the Hydrogen ions present.

### 3.1.1 Activity Coefficients

We have now seen and used several expressions for the chemical potential of a substance or component in a mixture. For a one-component ideal gas we had

$$\mu = \mu^\circ + RT \ln \frac{P}{P^\circ},$$

where $\mu^\circ$ is the chemical potential when $p = p^\circ$ and $p^\circ$ is usually one atmosphere.

For a mixture of ideal gases it can be shown that for each component, $i$, the chemical potential is given by,

$$\mu = \mu_i^o + RT \ln \frac{P}{P^\circ}.$$

For a non-ideal gas we used the fugacity, $f$, instead of the pressure and the chemical potential for one component is,

$$\mu = \mu^\circ + RT \ln \frac{f(P)}{P^\circ}$$

For a mixture of non-ideal gases it can be shown that,

$$\mu_i = \mu_i^o + RT \ln \frac{f_i}{P^\circ}$$

only now the fugacity of component $i$ is a function of the pressures of all the gases in the mixture,

$$f_i = f_i(p_1, p_2, p_3, \ldots).$$

For ideal solutions we found that,

$$\mu_i = \mu_i^o + RT \ln X_i.$$

These expressions for chemical potential all have the form of a reference or standard state chemical potential plus $RT$ times the logarithm of something related to pressure or concentration. This form turns out to be very important, so important that G. N. Lewis used it to give the most general case of chemical potential as

$$\mu_i = \mu_i^o + RT \ln a_i$$

The quantity, $a_i$, is called the "activity" of component $i$ and below equation is regarded as definition of activity.

All of the special cases we have been considering so far can be reconciled to this definition of activity. Thus, for an ideal gas mixture,

$$a_i = \frac{p_i}{p^\circ};$$

for a non-ideal gas mixture,

$$a_i = \frac{f_i}{p^\circ};$$

for an ideal solution,

$$a_i = X_i,$$

and so on.

In a non-ideal solution we would have to just write Equation 7 again,

$$\mu_i = \mu_i^o + RT \ln a_i.$$

In above equation, all non-idealities of the solution are absorbed into activity. We see that activity of a component of non-ideal solution from measurements of vapor pressure of the component in vapor in equilibrium with solution.

It is observed that chemical potential of a component must be the same in the vapor as in the liquid. So from above equations we obtain,

$$\mu_{il} = \mu_{ig}$$

$$\mu_{il}^* + RT \ln a_i = \mu_{ig}^o + RT \ln \frac{p_i}{p^\circ},$$

but for pure component $i$, we must have,

$$\mu_{il}^* = \mu_{ig}^o + RT \ln \frac{p_i^*}{p^\circ}.$$

When dealing with liquid solutions it is customary to write the chemical potential of the pure liquid as $\mu^*_{il}$ instead of the usual $\mu^\circ_{il}$, which means that the standard state for a liquid is the pure liquid itself. $p_i^*$ is the vapor pressure of the pure liquid. Subtracting above equations will give,

$$RT \ln a_i = RT \ln \frac{p_i}{p^\circ} - RT \ln \frac{p_i^*}{p^\circ}$$

$$= RT \ln \frac{p_i}{p_i^*}$$

hence

$$a_i = \frac{p_i}{p_i^*}$$

If the solution were ideal $p_i$ would be given by Raoult's law and the activity would be just the mole fraction. Further, it is easy to write the activity as product of an ideal part times a non ideality correction part.

For example, in a real solution we might write,

$$a_i = X_i \gamma_i,$$

or, in a real gas we would write,

$$f_i = p_i \gamma_i,$$

so that

$$a_i = \frac{f_i}{p^\circ} = \frac{p_i \gamma_i}{p^\circ} = \frac{p_i}{p^\circ} \gamma_i,$$

In the latter case we already know that

$$f(p) = pe \int_o^p \frac{(Z-1)}{p'} dp',$$

so that

$$\gamma = e \int_o^p \frac{(Z-1)}{p'} dp',$$

In case of non ideal gas $\gamma \to 1$ as $p \to 0$ while in case of non-ideal solution $\gamma_i \to 1$ as $X_i \to 1$. From above equations,

$$a_i = \frac{p_i}{p_i^*} = X_i \gamma_i,$$

from which we conclude that for a non-ideal solution,

$$\gamma_i = \frac{p_i}{X_i p_i^*} = \frac{\text{actual vapor pressure}}{\text{Raoult's law vapor pressure}}$$

In general, the activity coefficient is a unit less parameter that contains all of the non-ideality of a system.

### 3.1.2 Relationship between $K_a$ and $K_m$

The equilibrium constant given in terms of activities is the equilibrium constant. Using an example of dissociation equilibrium of a weak acid to show this relationship;

$$HA \to H^+ + A^-$$

equilibrium constant of reaction, (standard concentration, $m^\circ$)

$$K_a = \frac{a_+ a_-}{a_{HA}} = \frac{m_+ \gamma_+ m_- \gamma_-}{m_{HA} \cdot \gamma_{HA}} \cdot \frac{1}{m^o}$$

Knowing

$$\gamma_+ \cdot \gamma_- = \gamma_\pm^2$$

and $m_+ = m_- = m$

$$K_a = \frac{a_+ a_-}{a_{HA}} = \frac{m^2 \cdot \gamma_\pm^2}{m^o \cdot m_{HA} \cdot \gamma_{HA}}$$

$$K_a = K_c \cdot K_\gamma$$

where

$$K_c = \frac{m^2}{m^o \cdot m_{HA}}$$

$$K_\gamma = \frac{\gamma_\pm^2}{\gamma_{HA}}$$

The equilibrium constants given in terms of molality and activity coefficient $K_a$ = f($T$,$p$) but independent of molalities.

## 3.2 Debye-Hückel Limiting Law

Theoretical calculation of $\gamma_\pm$ will result from Debye Hückel limiting law which shows $\gamma_\pm$ in terms of *ionic strength I*, defined as,

$$I = \frac{1}{2}\sum_i m_i z_i^2$$

where

$z_i$ = charge on ion $i$

$m_i$ = molality of ion $i$

The ionic strength of solution is a measure of the amount of ions present which is a measure of the total concentration of charge in the solution.

A divalent ion (2+ or 2- ion, like $Ca^{2+}$) involves more in making solution ionic as compared to monovalent ion ($Na^+$). The ionic strength, emphasizes the charges of ions because the charge numbers occur as their squares.

**Example**

What is the value of ionic strength of HCl solution with molality 0.010:

$m_{H+} = m_{Cl-}$ = 0.01 mol/kg, and $z_{H^+}^2 = z_{Cl^-}^2 = 1$

$$I = \frac{1}{2}\left(0.01 \cdot (+1)^2 + 0.01 \cdot (-1)^2\right) = 0.01\ \text{mol} \cdot \text{kg}^{-1}$$

**Solution**

It is observed that for a simple salt of two monovalent ions, ionic strength is the concentration of the salt. Considering ionic strength of 0.10 molal $Na_2SO_4$

$$I = \frac{1}{2}\left(0.20(+1)^2 + 0.10(-2)^2\right)$$

$$= \frac{1}{2}(0.20 + .40)$$

= 0.30 molal

### 3.2.1 Debye-Hückel Theory

The chemical potential or activity of ions cannot be determined by pure thermodynamic. It results due to fact that the effects of an ion cannot be separated from the effects of the accompanying counter ion or electrochemical potential of ion cannot be separated into the chemical and electrical component. Such a separation must necessarily be based on a non-thermodynamic convention. The mean activity coefficient can be obtained at very low concentrations by Debye-Hückel Limiting Law *as*

$$\lg \gamma_{\pm} = -A \cdot |z_+ \cdot z_-| \cdot I^{1/2}$$

where

$A$ = 0.509 / (mol kg$^{-1}$)$^{1/2}$ for aqueous solution at 25 °C where A depends on relative permittivity of solvent and temperature.

The $\log_{10}$ of mean activity coefficient also depends on product of

$$|z_+ \cdot z_-|$$

the absolute value of cation and anion charge number.

When the ionic strength of the solution is too high for the limiting law to be valid, it is found that the activity coefficient may be estimated from the extended Debye-Hückel law.

$$\lg \gamma_{\pm} = \frac{-0.511 \cdot |z_+ \cdot z_-| I^{1/2}}{1 + b \cdot I^{1/2}}$$

Where $b$ is a measure of distance between ions.

In the limit of small concentration $I^{1/2} \ll 1$, and in the denominator of equation the term $b \cdot I^{1/2}$ can be neglected, hence as per Debye-Hückel Limiting Law:

$$\gamma_{\pm} = 10^{-0.511 \cdot A \cdot B}$$

where $A = |z_{+} \cdot z_{-}|; B = \dfrac{I^{1/2}}{1+I^{1/2}}$

Greater charges produce more negative the argument for the exponential, $\gamma_{\pm}$ is farther away from 1, its ideal value. The term $B$ varies between 0 and 1 when $I$ is small $B \sim 0$, and $\gamma_{\pm} \sim 1$.

## 3.3 Effect of Ionic Strength

The idea of ionic strength was introduced by Lewis and Randall while describing activity coefficients of strong electrolytes. Ionic strength of a solution is a measure of concentration of ions in that solution. Ionic compounds, when dissolved in water, dissociate into ions. The total electrolyte concentration in solution will affect important properties such as the dissociation constant or the solubility of different salts.

The main characteristics of solution with dissolved ions is ionic strength. Ionic strength can be molar (mol/L) or molal (mol/kg water) and to avoid confusion the units should be stated explicitly.

The ionic strength plays an important role in Debye Hückel theory which shows that the strong deviations from ideality typically encountered in ionic solutions. It is also important for the theory of double layer and related electro-kinetic phenomena and electro-acoustic phenomena in colloids and other heterogeneous systems.

Debye length, which is the inverse of Debye parameter ($\kappa$), is inversely proportional to the square root of ionic strength. Both molar and molal ionic strength have been used, often without explicit definition. Debye length is characteristic of the double layer thickness. Increasing the concentration or valence of the counter-ions compresses the double layer and increases the electrical potential gradient.

Media of high ionic strength are used in stability constant determination in order to minimize changes, during a titration, in the activity quotient of solutes at lower concentrations. Natural waters such as mineral water and seawater have often a non-negligible ionic strength due to the presence of dissolved salts which significantly affects their properties.

## 3.4 Activity Effects

### 3.4.1 Ionic Strength

In dealing with ionic solutions, it is known about ionic strength as it is observed that the ionic strength affects the activity of ions. For comparing experimental results, while working with solutions, ionic strength needs to be compared that shows interactions of ions with water molecules and other ions in a solution. This quantity is usually represented by I, and an explicit form will be given after we have defined the system.

Ionic strength is a measure of the concentration of the in a solution. It is typically expressed as the average electrostatic interactions among the ions in the electrolyte, and is half of the total obtained by multiplying the molality (the amount of substance per unit mass of solvent) of each ion by its valence squared.

Ionic strength is closely related to the concentration of electrolytes and indicates how effectively the charge on a particular ion is shielded or stabilized by other ions in an electrolyte. The main difference between ionic strength and electrolyte concentration is that the former is higher if some of the ions are more highly charged.

For instance, a solution of dissociated magnesium sulfate ($Mg^{+2}$ $SO_4^{-2}$) has 4 times higher ionic strength than solution of sodium chloride ($Na^{+}Cl^{-}$) with similar concentration. Also, ionic strength reflects the concentration of free ions, not by quantity of salt added to a solution, but by observing the salt dissolved with respective ions that bound together pair-wise, resembling uncharged molecules in solution. In this case the ionic strength is much lower than the salt concentration.

Ionic strength is an important factor in biochemical reactions and plays an essential role in the function of all living things. In order to survive, every organism must undergo a continuous series of chemical reactions that govern processes ranging from metabolism to respiration. Ionic strength is a key factor in these reactions because it affects the rates at which ions react with each other and, thus, the extent to which the reaction occurs.

In case of a person running, the moisture and electrolytes gets lost by way of perspiration. If the runner does not replace the lost electrolytes, then in such case the runner will become thirsty, sluggish, and overheated. Chemically, ionic strength of runner's remaining electrolytes would be very high as the essential chemical reactions in runner's body are affected by ionic strength, where the balance between the chemical reactions of life gets disrupted and the runner would get sick. Ionic strength is one of the basic characteristics of an organism's chemical makeup that determines whether that organism can exist in a state of homeostatis, or internal stability.

In case of acetylcholine, a positively charged ion which organisms releases at the ends of certain neurons where the purpose of acetylcholine is to serve as bridge among neurons, that passes along nerve impulses from one to other and to start muscular contractions. Ionic strength determines the rate at which acetylcholine reacts with other chemicals in an organism; therefore, if the ionic strength of the organism's electrolytes is too high, the acetylcholine will react at a rate too slow or too fast, or may bind too strongly or too weakly to its receptor for the organism to function normally.

Ionic strength is a also a useful parameter in the laboratory. The ionic strength of an electrolyte can tell chemists a great deal about the dynamics of specific chemical reactions. Scientists can also change an electrolyte's ionic strength to alter the outcome of a reaction in a measured and deliberate way.

High ionic strength chloride solutions are useful in many geological, environmental and industrial processes. For example, oil companies are researching the use of high ionic strength solutions to reduce the cost of controlling scale formation in pipes; the metallurgical industry uses them to improve the efficiency of copper extraction from chalcopyrite ores; and the nuclear waste industry uses them to understand the behavior of radionuclides in underground storage deposits where there are salt deposits. Because scientists do yet not fully understand the theoretical behavior of high ionic strength chloride solutions systems, most applications are based on observed models.

The dissociation of an electrolyte $M_xX_m$ is,

$$M_xX_m = x\ M^{m+} + m\ X^{x-}$$

Positive ions $M^{m+}$ and negative ions $X^{x-}$ must be present together in one solution, and there is no way to separate activities of positive and negative ions. Thus, we usually use a mean ionic activity (a,

$$a = g\ C\ (m^m x^x)^{-(m+x)}$$

for both positive and negative ions, C being the stoichiometric concentration of the electrolyte. The Debye-Huckel limiting law shows that the activity coefficient is related to the ionic strength, I in the following way:

$$\ln g = -\ A\ z_+ z_- I^{1/2}$$

where A is a constant (= 1.172 at 298 K), $z_+ z_-$ the valence factor and I is the ionic strength which is define by the equation:

$$I = ½\ S\ z_i^2 m_i$$

where $m_i$ is the concentration of the ith ion concentration. The summation, S, is taken over all the possible ions in the solution.

**Example**

What is the ionic strength for a 1.0 M NaCl solution?

**Solution**

Using the simple formula for ionic strength *I* given above, we have

$$I = ½\ (1*1^2 + 1*1^2)$$

$$= 1.00 \text{ (a unitless quantity)}$$

**Example**

What is the ionic strength for a solution whose concentrations are 1.0 M $La_2(SO_4)_3$ plus 1.0 M $CaCl_2$?

**Solution**

For this solution, the concentrations are:

$[La^{3+}] = 2.0$ M

$[SO_4^{2-}] = 3.0$ M

$[Ca^{2+}] = 1.0$ M

$[Cl^-] = 2.0$ M

$I = ½ (2*3^2 + 3*2^2 + 1*2^2 + 2*1^2)$

$= 18.0$

**Example**

Estimate the activity coefficient for $Na^+$ in a solution whose ionic strength is 0.01 at 298 K.

**Solution**

Using the limiting Debye-Huckel law,

$\ln g = - A z_+ z_- I^{½}$

$= -1.172 * 1 * (0.01)^{-½}$

$= -0.1172$

$g = 0.90$

## 3.5 Chemical Activity Coefficients of Ions

Activity of a molecule or ion in a solution is referred to **effective concentration.** From the definition of activity from equilibrium thermodynamics, we have

$$a = \gamma \frac{m_i}{m_0}$$

$m_0$ = standard concentration, (e.g.: 1 mol $kg^{-1}$),

$m_i$ = molality of $i^{th}$ component of a solution,

$\gamma$ = concentration dependent activity coefficient. Both a and $\gamma$ have no unit.

On diluting the solution, i.e.

$c \rightarrow 0$

$$\gamma \to 1$$

then $a = c$

activity becomes equal to the concentration. In very dilute solution interactions are insignificant and concentration and effective concentration are the same.

Though, activity and activity coefficient have no unit their magnitude depend on the concentration units used. E.g., concentration of the same solution can be given in units $mol \cdot kg^{-1}$, $mol \cdot dm^{-3}$, molar fraction and so on.

Activities of ionic solutions serve us for calculating

- accurate chemical potentials
- accurate equilibrium constants

Taking into account that the activity has a greater importance in ionic solution than that of non-electrolytes. Interaction forces are greater among ions (Coulombic interaction).

### 3.5.1 Ions in Aqueous Solution

Strong electrolytes are the ionic compounds that completely dissociate into ions when they are dissolved.

$$NaCl(aq) \rightarrow Na^{+}(aq) + Cl^{-}(aq).$$

Weak electrolytes are ionic compounds that partially dissociate. The chemical potentials for the formation of ions in aqueous solution

$$\mu_{NaCl,\,aq} = \mu_{Na^{+},\,aq} + \mu_{Cl^{-},\,aq}$$

which also implies that standard chemical potentials

$$\mu^{0}_{NaCl,\,aq} = \mu^{0}_{Na^{+},\,aq} + \mu^{0}_{Cl^{-},\,aq}$$

We also know that we can write Equation 2. again with standard chemical potential and activity

$$\mu_{NaCl,\,aq} = \mu^{0}_{NaCl,\,aq} + RT \ln a_{NaCl,\,aq}$$

When working with ionic solutions we would like to be a little more specific about the activities of the species in solution. It is customary to use units of molality, m, for ions and compounds in aqueous solution. If the solutions were ideal we could write for the activity of $i^{th}$ component

$$a_{i} = \frac{m_{i}}{m_{0}}$$

and the chemical potential would be written

$$\mu_i = \mu_i^0 + RT \ln \frac{m_i}{m_i^0}$$

Two things must be said about equation, that it must be understood that there is an implied $m_i^o$ dividing the $m_i$ inside the logarithm, and standard state is the solution at concentration $m_i = m_i^o$. Usually we set $m_i^o = 1$ mol kg$^{-1}$. However, ionic solutions are far from ideal so we must correct this expression for chemical potential for the non-idealities. As usual, we will use an activity coefficient, $\gamma$ , and write the activity as

$$a_i = \gamma_i \cdot \frac{m_i}{m_0}$$

and consistent with what we have been doing, we will set $m_i^o$ to 1 molal and not write it in the equation. The form of equation indicates that activity and activity coefficient are dimensionless, i.e. they have no unit. Thus the chemical potential will be written

$$\mu_i = \mu_i^0 + RT \ln m_i \gamma_i$$

The standard state for this equation is a hypothetical standard state. The standard state is not actually realizable. We are using the so-called Henry's law standard state in which the solution obey's Henry's law in the limit of infinite dilution. That is,

$\gamma_i \rightarrow 1$ in the limit when $m_i \rightarrow 0$.

We will simplify matters by writing $m_+$ , $\gamma_+$ and $m_-$ ,$\gamma_-$ for the molalities and activity coefficients of the positive and negative ionic species, respectively.

We will also refer to the ionic compound simply as the "salt." With this notation we can rewrite former Equations as,

$$\mu_{\text{salt}} = \mu_+ + \mu_-$$

$$or\, \mu_{\text{salt}}^0 + RT \ln a_{\text{salt}} = \mu_+^0 + RT \ln m_+ \gamma_+ + \mu_-^0 + RT \ln m_- \gamma_-$$

In a simplified form

$$\mu_{\text{salt}}^0 + RT \ln a_{\text{salt}} = \mu_+^0 + \mu_-^0 + RT \ln m_+ \gamma_+ \cdot m_- \gamma_-$$

The sum of standard chemical potentials is equal to that of the salt

$$\mu_{\text{salt}}^0 = \mu_+^0 + \mu_-^0$$

The logarithmic term in equation shown on left is identical to logarithmic term on right,

$$RT \ln a_{\text{salt}} = RT \ln m_+\gamma_+ \cdot m_-\gamma_-$$

from which we conclude that

$$a_{\text{salt}} = m_+\gamma_+ \cdot m_-\gamma_-$$

The activity coefficients, $\gamma_+$ and $\gamma_-$ can't be measured independently because solutions must be electrically neutral. In other words, you can't make a solution which has just positive or just negative ions. You can't calculate the individual activity coefficients from theory, either. However, you can measure a "geometric mean" activity coefficient and, within limits, you can calculate it from theory.

## 3.6 Water Activity

Water activity ($a_w$) describes as equilibrium, which is the amount of water available for hydration of materials where water activity = water availability that serves as ratio of the thermodynamic activity to analytical concentration. When it is available, $a_w = 1$, and when it is not there, $a_w = 0$ . If water interacts with solutes and surfaces, it is unavailable for other hydration interactions and its effective concentration in the solution is reduced.

Water activity ($a_w$) is one of the most critical factors in determining quality and safety of the goods you consume every day. Water activity affects the shelf life, safety, texture, flavor, and smell of foods. It is also important to the stability of pharmaceuticals and cosmetics. While temperature, pH and several other factors can influence if and how fast organisms will grow in a product, water activity may be the most important factor in controlling spoilage. Most bacteria, for example, do not grow at water activities below 0.91, and most molds cease to grow at water activities below 0.80.

By measuring water activity, it is possible to predict which microorganisms will and will not be potential sources of spoilage. Water activity–not water content–determines the lower limit of available water for microbial growth. In addition to influencing microbial spoilage, water activity can play a significant role in determining the activity of enzymes and vitamins in foods and can have a major impact their color, taste, and aroma. It can also significantly impact the potency and consistency of pharmaceuticals.

A water activity value of unity indicates pure water whereas zero indicates the total absence of 'free' water molecules; addition of solutes always lowering the water activity. Unity water activity does not imply the bulk water is unstructured, only that it is indistinguishable from 'pure' water.. Water activity has been reviewed in aqueous and biological systems and has particular relevance in food chemistry and preservation.

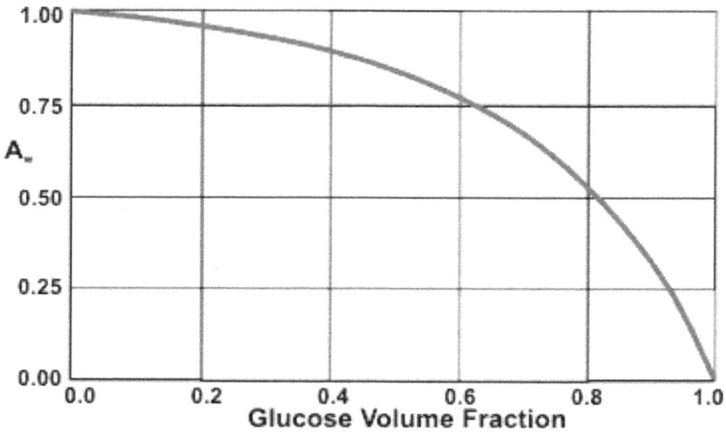


**Fig.:** Water activity of glucose

Water activity ($a_w$) is defined as ratio of the fugacity (f) of the water to its fugacity under reference conditions ($f_0$), but it approximates well to the more easily determined ratio of partial pressures (*P*) under normal working conditions.

$$a_w = f/f_0 \sim P/P_0$$

The activity coefficient ($\gamma_w$) has dependence on partial molar volume and hydrogen bond strength that covers dependence on temperature and dielectric constant of water and in dilute solutions which can be approximated by unity. The water activity ($a_w$) is related to the chemical potential ($\mu_w$; at equilibrium, $\mu_w$ of liquid water and its vapor phase are identical) by

$$\mu_w = \mu_w^\circ + RTLn(a_w)$$

where $\mu_w^\circ$ is the standard chemical potential of water.

At equilibrium $\mu_w$ is the same everywhere in the system establishing this equilibrium by movement of water from regions of high water activity to regions of low water activity. Multiplication of water activity by 100 gives the equilibrium relative humidity (ERH) in percent.

$$a_w = P/P_0 = \text{ERH (\%)} / 100$$

Thus, water activity is easily and accurately determined from equilibrium relative humidity._Water activity is the effective mole fraction of water, defined as:

$$a_w = \gamma_w x_w = P/P_0$$

where

$\gamma_{w=}$ activity coefficient of water

$x_{w=}$ mole fraction of water in aqueous fraction

$P$ = partial pressure of water above material

$P_{0}$ = partial pressure of pure water at same temperature

It may be experimentally determined from the dew-point temperature of the atmosphere in equilibrium with the material; for example, by use of a chilled mirror (in a hygrometer) to show the temperature when the air becomes saturated in equilibrium with water. A high $a_w$ (that is, > 0.8) indicates a 'moist' or 'wet' system and a low $a_w$ (that is, < 0.7) generally indicates a 'dry' system. Water activity reflects a combination of water-solute and water-surface interactions plus capillary forces. The nature of a hydrocolloid or protein polymer network can thus affect the water activity, cross linking reducing the activity. Note that the water activity of any aqueous solution in equilibrium with ice ($a_w^{\,i}$) is equal to the water vapor pressure over ice to the water pressure over pure liquid water and does not depend on the solute's nature or concentration. Solutions with the same ice melting point therefore have the same water activity.

Water activity (aw) refers to the free or available water within a food which is not bound to food molecules, i.e. water which is available in foods to support the growth of bacteria, mold and yeast. Water activity has a scale from zero, i.e. (0% water available) to one (100% water available).

Routine water activity measurement of food products is not as common place as pH measurement in the food industry. However, with advances in instrumentation and methodology, its use is becoming more widespread. Typically, water activity is measured on macerated food. It is recommended that food business operators measure water activity according to EN ISO 21807:199416 or other applicable methods.

Water activity is derived from fundamental principles of thermodynamics and physical chemistry. As a thermodynamic principle there are requirements in defining water activity that must be met. These requirements are: pure water ($a_{w=}$ 1.0) is the standard state, the system is in equilibrium, and the temperature is defined.

In the equilibrium state:

$$\mu = \mu_o + RT \ln(f/f_o)$$

where: $\mu$ (J mol$^{-1}$) is the chemical potential of the system i.e. thermodynamic activity or energy per mole of substance; $\mu_o$ is the chemical potential of the pure material at the temperature T (°K); R is the gas constant (8.314 J mol$^{-1}$ K$^{-1}$) ; f is the fugacity or the escaping tendency of a substance; and $f_o$ is escaping tendency of pure material. The activity of a species is defined as $a = f/f_o$. When dealing with water, a subscript is designated for the substance,

$$a_w = f/f_o$$

$a_w$ is activity of water, or the escaping tendency of water in system divided by the escaping tendency of pure water with no radius of curvature. For practical purposes, under most conditions in which foods are found, the fugacity is closely approximated by the vapor pressure (f ~ p) so;

$$a_w = f/f_o \sim p/p_o$$

Equilibrium is obtained in a system when μ is the same everywhere in the system. Equilibrium between the liquid and the vapor phases implies that μ is the same in both phases. It is this fact that allows the measurement of the vapor phase to determine the water activity of the sample.

Water activity is defined as the ratio of the vapor pressure of water in a material (p) to the vapor pressure of pure water ($p_o$) at the same temperature. Relative humidity of air is defined as the ratio of the vapor pressure of air to its saturation vapor pressure. When vapor and temperature equilibrium are obtained, the water activity of the sample is equal to the relative humidity of air surrounding the sample in a sealed measurement chamber. Multiplication of water activity by 100 gives the equilibrium relative humidity (ERH) in percent.

$$a_w = p/p_o = \text{ERH (\%)} / 100$$

Water activity is a measure of the energy status of the water in a system. There are several factors that control water activity in a system. Colligative effects of dissolved species (e.g. salt or sugar) interact with water through dipole-dipole, ionic, and hydrogen bonds. Capillary effect where the vapor pressure of water above a curved liquid meniscus is less than that of pure water because of changes in the hydrogen bonding between water molecules. Surface interactions in which water interacts directly with chemical groups on undissolved ingredients (e.g. starches and proteins) through dipole-dipole forces, ionic bonds ($H_3O$+ or OH-), van der Waals forces (hydrophobic bonds), and hydrogen bonds. It is a combination of these three factors in a food product that reduces the energy of the water and thus reduces the relative humidity as compared to pure water. These factors can be grouped under two broad categories osmotic and matric effects.

## 3.7 Kinetic Salt Effect

Kinetic salt effect shows the variation of rate constant of a reaction with respect to ionic strength of the solution. Mathematically, consider the ionic strength of the solution, which has no effect of ionic radius, the kinetic salt effect describes the way the salt stabilizes the reactants.

For example, in the reaction, I−+C+⇌ Eo → Product, where each reactant has a charge. The negatively-charged reactant is stabilized by the positive charges from the salt, and the positively-charged reactant is stabilized by the negative charges from the salt. As a result, the rate at which the reactants come together decreases, thus decreasing the rate at which E forms. Because a charged intermediate is also stabilized in the solution, the half life of the intermediate at equilibrium increases, shifting the reaction toward product formation.

The kinetic salt effect is the primary which is purely kinetic origin that depends on the activity coefficients, or can be secondary, which cause a change of concentration of reacting ions or molecules, due to a static salt effect.

# 4 Strong Acids and Bases

## 4.1 Nature of Acids and Bases

Acids and bases were first described and categorized by Swedish chemist, Svante Arrhenius in the late 19th century. Arrhenius, being the first to describe acids and bases, identified what are now known as the strong acids and strong bases.

Matters can be classified in many ways. Acids and bases are another way of classification of matters. Most of the reactions taking place in water solutions are in acid or base mediums. Definitions of acids and bases are given by Arrhenius and Bronsted -Lowery. They define acids and bases in different ways. Now we examine these definitions by one.

### 4.1.1 Definition of Arrhenius

Arrhenius defines acids as " in water solutions matters that give $H^+$ ion are called **acid**". Examples of acids are;

$$HCl(aq) \rightarrow H^+(aq) + Cl^-$$

$$HNO_3(aq) \rightarrow H^+(aq) + NO_3^-(aq)$$

$$H_3PO_4(aq) \rightarrow 3H^+(aq) + PO_4^{-3}(aq)$$

$$H_2SO_4(aq) \rightarrow 2H^+(aq) + SO_4^{-2}(aq)$$

As you can see from the examples, acids give $H^+$ ion to solution. On the contrary, some of the acids like $SO_2$, $P_2O_5$, $SO_3$ do not include H atom, but their solutions with water shows acidic property. Solutions of these matters with water becomes;

$$SO_2(g) + H_2O(l) \rightarrow H_2SO_3(aq)$$

$$CO_2(g) + H_2O(l) \rightarrow H_2CO_3(aq)$$

$$P_2O_5(s) + 3H_2O(l) \rightarrow 2H_3PO_4(aq)$$

These non metal oxides are called "**anhydrous acids**".

Arrhenius defines bases as " in water solutions matters that give $OH^-$ ion are called **bases**". Examples of bases are;

$$KOH(s) \rightarrow K^+(aq) + OH^-(aq)$$

$$Ba(OH)_2(s) \rightarrow Ba^{+2}(aq) + 2OH^-(aq)$$

$$NaOH(s) \rightarrow Na^+(aq) + OH^-(aq)$$

$$Ca(OH)_2(s) \rightarrow Ca^{+2}(aq) + 2OH^-(aq)$$

As you can see from the examples, bases give $OH^-$ ion to solution. On the contrary, some of the bases like $NH_3$, MgO, CaO do not include $OH^-$ ion, but their solutions with water includes $OH^-$ ion and they show basic property. Look at these examples given below;

$$NH_3(aq) + H_2O(l) \rightarrow NH_4^+(aq) + OH^-(aq)$$

$$K_2O(s) + H_2O(l) \rightarrow 2K^+(aq) + 2OH^-(aq)$$

$$MgO(s) + H_2O(l) \rightarrow Mg^{+2}(aq) + 2OH^-(aq)$$

These matters are called "**anhydrous bases**".

### 4.1.2 Features of Acids

Acids can be defined in many ways but generally their aqueous solutions have the following properties.

They are sour in taste.

They turn blue litmus red.

They react with certain metals and liberate hydrogen gas.

They react with oxides and hydroxides of metals forming salt and water.

Their aqueous solutions conduct electricity

**Example**

$$CaCO_3 + 2HNO_3 \rightarrow Ca(NO_3)_2 + CO_2(g) + H_2O$$

$$NaHCO_3 + HCl \rightarrow NaCl + CO_2(g) + H_2O$$

They react with active metals and produce salt and $H_2$ gas.

**Example**

$$Zn + H_2SO_4(sol) \rightarrow ZnSO_4(sol) + H_2(g)$$

$$Mg + 2HCl \rightarrow MgCl_2 + H_2$$

$$2Al + 3H_2SO_4 \rightarrow Al_2(SO_4)_3 + 3H_2$$

Some of the metals like Au, Pt, Ag, Cu and Hg are exceptions of this property. They are called noble metals. They do not form $H_2$ gas in reactions with acids. However, some of noble metals react with $HNO_3$ and $H_2SO_4$ and do not produce $H_2$.

**Example**

$$Cu + HCl \rightarrow \text{No reaction occurrs}$$

$$Cu + 2H_2SO_4 \rightarrow CuSO_4 + SO_2 + 2H_2O$$

$$3Cu + 8HNO_3 \rightarrow 3Cu(NO_3)_2 + 2NO + 4H_2O$$

Acids turn blue litmus to red.

Acids react with bases and form salt and water. This reaction type is called **neutralization reaction.**

$$H_2SO_4 + 2NaOH \rightarrow Na_2SO_4 + 2H_2O$$

$$2HCl + Ba(OH)_2 \rightarrow BaCl_2 + 2H_2O$$

$$HCl + NaOH \rightarrow NaCl + H_2O$$

### 4.1.3 Key Features of Bases

- Bases are defined in various ways but generally substances having the following characteristics are called bases.
- They have a bitter taste.
- Their aqueous solutions have a soapy touch.
- They turn red litmus blue.
- They react with acids to form salt and water.
- Their aqueous solutions conduct electricity.

**Example**

$$2Al + 6NaOH \rightarrow 2Na_3AlO_3 + 3H_2$$

$$Zn + 2NaOH \rightarrow Na_2ZnO_2 + H_2$$

They react with acids and form salt and water. (Neutralization reactions)

## 4.2 Definition of Bronsted-Lowry

There are some limitations in Arrhenius definition. For example, it can not explain anhydrous acids and bases. Thus, Bronsted-Lowry stated another definition for acids and bases as;

Acids are matter that donates $H^+$ and bases are matters that accept $H^+$ ion.

$$HCl(g) + NH_3(g) \rightarrow NH_4^+(s) + Cl^-(s)$$

In this reaction, HCl donates $H^+$ ion so it is acid and $NH_3$ accepts $H^+$ ion, it is base.

$$CO_3^{-2} + H_2O \rightarrow HCO_3^- + OH^-$$

In this reaction one acid and one base react to give another acid and base. $H_2O$ acid donates $H^+$ ion and becomes $OH^-$ base, $CO_3^-$ base accepts $H^+$ ion and becomes $HCO_3^-$ acid. $OH^-$ is the **conjugate base** of $H_2O$ and $HCO_3^-$ is **conjugate acid** of $CO_3^{-2}$. List given below shows some common acid and base conjugates;

| Acid | Base |
|---|---|
| $H_2SO_4$ | $HSO_4^-$ |
| $HSO_4^-$ | $SO_4^{-2}$ |
| $H_3O^+$ | $H_2O$ |
| $H_2O$ | $OH^-$ |
| $H_2CO_3$ | $HCO_3^-$ |
| $HCO_3^-$ | $CO_3^{-2}$ |

**Example**

According to Bronsted-Lowry definition, which ones of the following statements are true for following reaction.

$$HCO_3^- + HSO_4^- \rightarrow H_2CO_3 + SO_4^{-2}$$

i) Acidic property of $HSO_4^-$ ion is higher than acidic property of $HCO_3^-$

ii) $SO_4^{-2}$ ion shows both acidic and basic property.

iii) Acidic property of $H_2CO_3$ is higher than $HSO_4^-$ ion.

**Solution**

In given reaction, $HSO_4^-$ and $H_2CO_3$ are acid and $HCO_3^-$ and $SO_4^{-2}$ are base. Since $HSO_4^-$ ion donate $H^+$ ion to $HCO_3^-$, acidic property of it is high. I is true.

Acids should have $H^+$ ion in its structure to become Bronsted-Lowry acid. Since there is no $H^+$ ion in $SO_4^{-2}$, it does not behave like acid. II is false.

Third statement can be true or false, we can not say it is true by looking at this reaction.

**Example**

Which ones of the following couples are acid in given reaction?

$$H_2PO_4^- + HSO_3^- \leftrightarrow HPO_4^- + H_2SO_3$$

i) $H_2PO_4^-$ and $H_2SO_3$

ii) $HSO_3^-$ and $H_2SO_3$

iii) $HSO_3^-$ and $HPO_4^-$

iv) $H_2PO_4^-$ and $HPO_4^-$

**Solution**

$H_2PO_4^-$ donates $H^+$ and becomes $HPO_4^-$ so it shows acidic property.

$H_2SO_3$ donates $H^+$ and becomes $HSO_3^-$ so it shows acidic property.

## 4.2.1 Limitations of Theory

Hydrochloric acid is neutralised by both sodium hydroxide solution and ammonia solution. In both cases, you get a colourless solution which you can crystallise to get a white salt - either sodium chloride or ammonium chloride. These are clearly very similar reactions. The full equations are:

$$NaOH_{(aq)} + HCl(aq) \rightarrow NaCk(aq) + H_2O_{(l)}$$

$$NH_{3(aq)} + HCl(aq) \rightarrow NH_4Cl(aq)$$

In the sodium hydroxide case, hydrogen ions from the acid are reacting with hydroxide ions from the sodium hydroxide - in line with the Arrhenius theory. However, in the ammonia case, there don't appear to be any hydroxide ions. But if you look at the equations carefully, the ammonia is in solution - $NH_{3(aq)}$.

Ammonia reacts with water as:

$$NH_{3(aq)} + H_2O_{(l)} \square \quad NH_4^+{}_{(aq)} + OH^-_{(aq)}$$

This is a reversible reaction, and in a typical dilute ammonia solution, about 99% of the ammonia remains as ammonia molecules. Nevertheless, there are hydroxide ions there and those react with hydrogen ions in just the same way as hydroxide ions from sodium hydroxide.

Now the ammonia can be considered as base on Arrhenius definition which generates hydroxide ions in solution. But many reactions are direct reaction which occurs between ammonia molecules and hydrogen ions - which doesn't fit the Arrhenius definition. This same reaction also happens between ammonia gas and hydrogen chloride gas.

$$NH_{3(g)} + HCl_{(g)} \rightarrow NH_4Cl_{(s)}$$

In this case, there aren't any hydrogen ions or hydroxide ions in solution - because there isn't any solution. The Arrhenius theory wouldn't count this as an acid-base reaction, despite the fact that it is producing the same product as when the two substances were in solution.

## 4.3 Brønsted-Lowry Conjugates Acid-Base Pairs

### 4.3.1 Theory

According to the theory, an acid is a proton which is a hydrogen ion donor while a base is a proton with hydrogen ion acceptor. As per Bronsted-Lowry which is not against the Arrhenius theory, but is the addition of the theory which says that a hydroxide ions are bases as they accept hydrogen ions from acids and form water. An acid produces hydrogen ions in solution because it reacts with the water molecules by giving a proton to them.

When hydrogen chloride gas dissolves in water to produce hydrochloric acid, the hydrogen chloride molecule gives a proton (a hydrogen ion) to a water molecule. A co-ordinate (dative covalent) bond is formed between one of the lone pairs on the oxygen and the hydrogen from the HCl. Hydroxonium ions, $H_3O^+$, are produced.

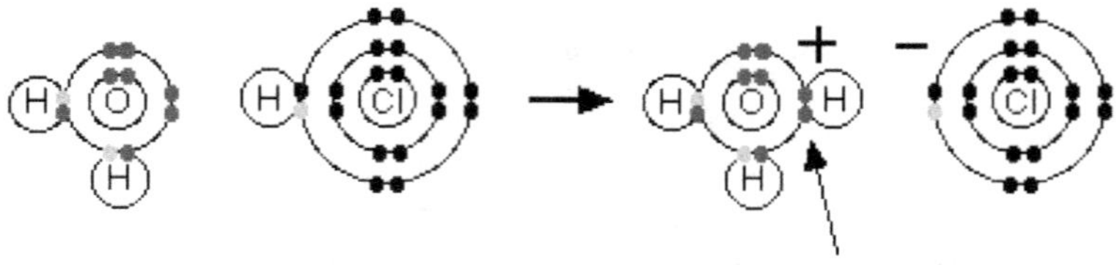


$$H_2O + HCl \rightarrow H_3O^+ + Cl^-$$

In case of reaction of acid in a solution with base, functioning of acid results in hydroxonium ion. In an example, a proton is transferred from a hydroxonium ion to a hydroxide ion to make water.

$$H_3O^+_{(ag)} + OH^-_{(aq)} \rightarrow 2H_2O_{(l)}$$

Showing the electrons, but leaving out the inner ones:

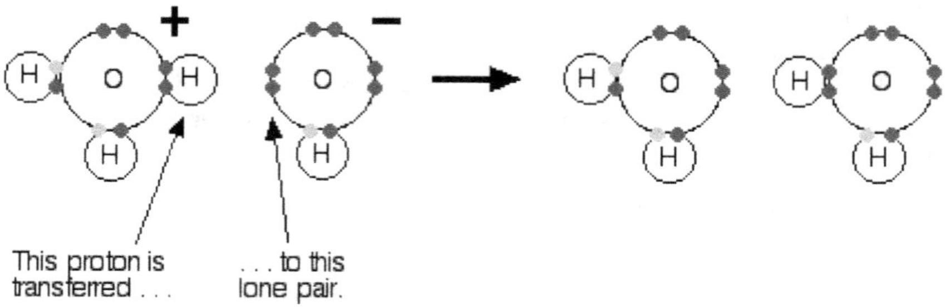


Hydroxonium Ions.

It is important to realise that whenever it is hydrogen ions in solution, $H^+_{(aq)}$ it is actually discussing about hydroxonium ions. As discussed by Brønsted, all acid-base reactions involves transfer of an $H^+$ ion or proton. Water reacts with itself, for example, by transferring an $H^+$ ion from one molecule to another to form an $H_3O^+$ ion and an $OH^-$ ion.

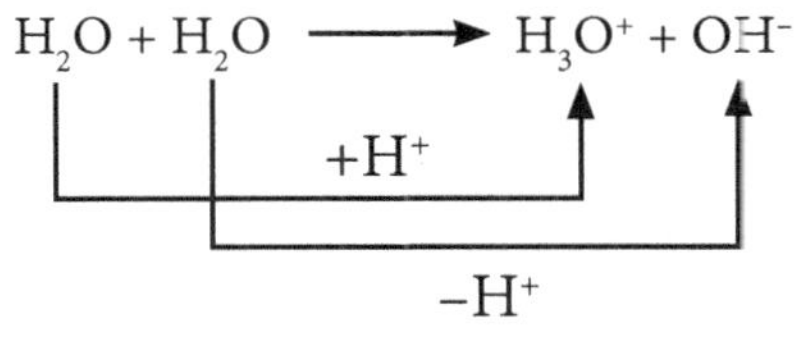


Reaction of Water.

As per the theory, acid is a proton donor while base is a proton acceptor. Acids are often divided into categories such as "strong" and "weak." One measure of the strength of an acid is the acid-dissociation equilibrium constant, $K_a$, for that acid.

$$K_a = \frac{\left[H_3O^+\right]\left[A^-\right]}{[HA]}$$

When $K_a$ is relatively large, we have a strong acid.

HCl: $K_a = 1 \times 10^3$

When it is small, we have a weak acid.

$CH_3CO_2H$: $K_a = 1.8 \times 10^{-5}$

When it is very small, we have a very weak acid.

$H_2O$: $K_a = 1.8 \times 10^{-16}$

As suggested that the enormous range of concentrations of $H_3O^+$ and $OH^-$ ions in aqueous solutions be compressed in more manageable set of data by taking advantage of logarithmic mathematics and calculating then the pH or pOH of the solution can be:

$$pH = -\log [H_3O^+]$$

$$pOH = -\log [OH^-]$$

The "p" in pH and pOH is an operator that indicates that the negative of the logarithm should be calculated for any quantity to which it is attached, thus, $pK_a$ is negative of logarithm of acid-dissociation equilibrium constant.

$$pK_a = -\log K_a$$

The only disadvantage of using $pK_a$ as a measure of the relative strengths of acids is the fact that large numbers now describe weak acids, and small (negative) numbers describe strong acids.

| | |
|---|---|
| 2HCl: | $pK_a$ = -3 |
| $CH_3CO_2H$: | $pK_a$ = 4.7 |
| $H_2O$: | $pK_a$ = 15.7 |

An important features of the Brønsted theory is the relationship it creates between acids and bases. Every Brønsted acid has a conjugate base, and vice versa.

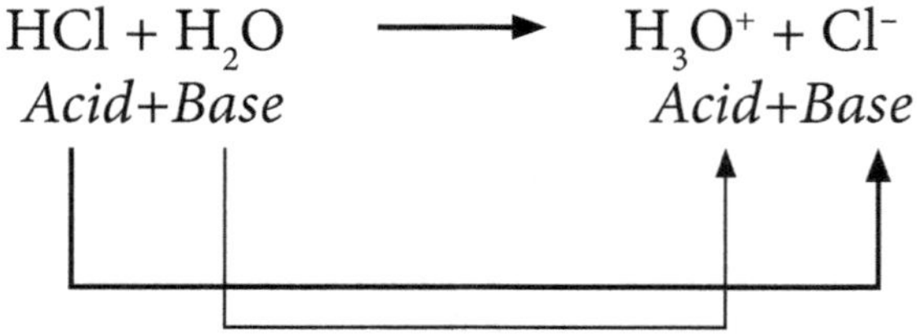


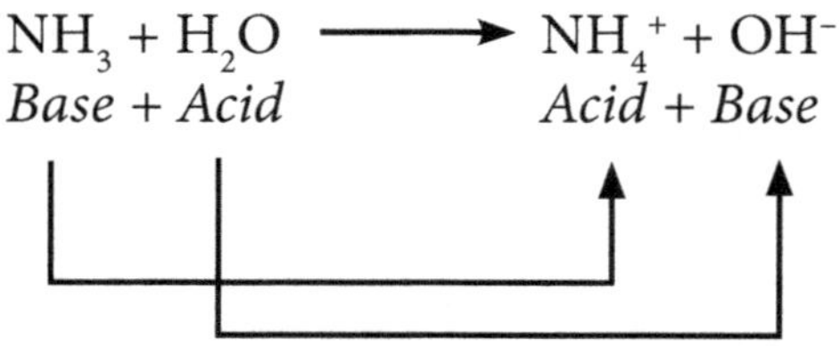


Just as the magnitude of $K_a$ is a measure of the strength of an acid, the value of $K_b$ reflects the strength of its conjugate base. Consider what happens when we multiply the $K_a$ expression for a generic acid (HA) by the $K_b$ expression for its conjugate base ($A^-$).

$$\frac{\left[H_3O^+\right]\left[A^-\right]}{[HA]}\times\frac{\left[HA\right]\left[OH^-\right]}{\left[A^-\right]}=\left[H_3O^+\right]\left[OH^-\right]$$

If we now replace each term in this equation by the appropriate equilibrium constant, we get the following equation.

$$K_aK_b = K_w = 1 \times 10^{-14}$$

Because the product of $K_a$ times $K_b$ is a relatively small number, either the acid or its conjugate base can be «strong.» But if one is strong, the other must be weak. Thus, a strong acid must have a weak conjugate base.

$$\underset{\substack{Strong\\acid}}{HCl}+H_2O\rightarrow H_3O^+ + \underset{\substack{Weak\\base}}{Cl^-}$$

A strong base, on the other hand, must have a weak conjugate acid.

$$NH_4^+ + \underset{\substack{Strong\\base}}{OH} \rightarrow NH_3 + \underset{\substack{Weak\\acid}}{H_2O}$$

### 4.3.2 Conjugate Pairs

When hydrogen chloride dissolves in water, almost 100% of it reacts with the water to produce hydroxonium ions and chloride ions. Hydrogen chloride is a strong acid, and we tend to write this as a one-way reaction:

$$H_2O + HCl \rightarrow H_3O^+ + Cl^-$$

In fact, the reaction between HCl and water is reversible, but only to a very minor extent. In order to generalise, consider an acid HA, and think of the reaction as being reversible.

$$HA + H_2O \rightleftharpoons H_3O^+ + A^-$$

Thinking about the forward reaction:

- The HA is an acid because it is donating a proton (hydrogen ion) to the water.
- The water is a base because it is accepting a proton from the HA.

But there is also a back reaction between the hydroxonium ion and the $A^-$ ion:

- The $H_3O^-$ is an acid because it is donating a proton (hydrogen ion) to the $A^-$ ion.
- The $A^-$ ion is a base because it is accepting a proton from the $H_3O^+$.

The reversible reaction contains *two* acids and *two* bases. We think of them in pairs, called conjugate pairs.

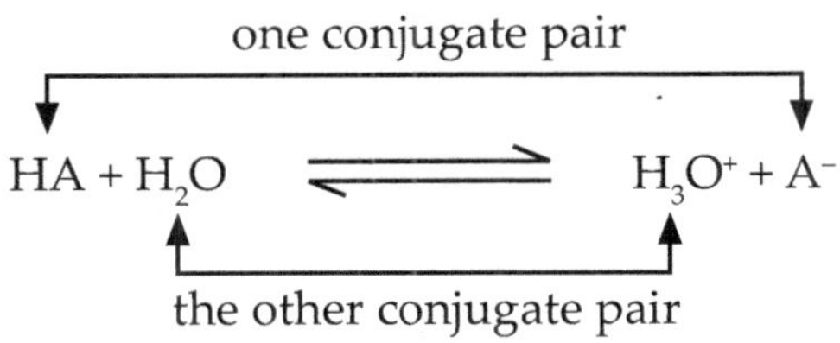


When the acid, HA loses a proton it forms a base $A^-$ while when the base $A^-$, accepts a proton back again, it forms the acid, HA. These two are a conjugate pair where members of a conjugate pair differ from each other by presence or absence of the transferable hydrogen ion.

If HA is acid, then $A^-$ is its conjugate base.

If $A^-$ is base, then HA is its conjugate acid.

The water and the hydroxonium ion are also a conjugate pair. Think of water as a base, the hydroxonium ion is its conjugate acid as it has extra hydrogen ion which can be given again. Think about hydroxonium ion as an acid, then water is its conjugate base as water can accept a hydrogen ion back again to reform hydroxonium ion.

### Strong Acid - Weak Conjugate Base Pair

A strong Brønsted-Lowry acid is one which has strong tendency to donate a proton, so strong acids include $H_3O^+$, HCl and $HNO_3$. For example, in water, strong acid like hydrochloric acid readily donates a proton to a water molecule:

| | | | | | | |
|---|---|---|---|---|---|---|
| Acid | + | Base (water) | ⇋ | Conjugate base of HCl acid | + | Conjugate acid of base $H_2O$ |
| HCl | + | $H_2O$ | ⇋ | $Cl^-$ | + | $H_3O^+$ |

The equilibrium position lies so far to the right that we usually assume that the HCl molecule completely dissociates in water:

| | | | | | | |
|---|---|---|---|---|---|---|
| Acid | + | Base (water) | → | Conjugate base of HCl acid | + | Conjugate acid of base $H_2O$ |
| HCl | + | $H_2O$ | → | $Cl^-$ | + | $H_3O^+$ |

This means that the reverse reaction does not occur to any appreciable extent, that is $Cl^-$ does not accept a proton from $H_3O^+$.

So $Cl^-$ must be a weak base, it has very little tendency to accept a proton.

The conjugate base of a strong acid is a weak base.

### Moderately Weak Acid - Moderately Weak Conjugate Base Pair

A moderately weak Brønsted-Lowry acid has only a slight tendency to donate a proton. Moderately weak acids include $CH_3COOH$, $HNO_2$, $H_2S$, $NH_4^+$, $HCO_3^-$

For example, acetic acid (ethanoic acid) is a moderately weak acid in aqueous solution. Some of the acetic acid (ethanoic acid) molecules dissociate in water, producing acetate ions (ethanoate ions) and protons.

Some of the acetate ions (ethanoate ions) and protons reform acetic acid (ethanoic acid), so that the system is in equilibrium:

| | | | | | | |
|---|---|---|---|---|---|---|
| Acid | + | Base (water) | ⇋ | Conjugate base of HCl acid | + | Conjugate acid of base $H_2O$ |
| $CH_3COOH$ | + | $H_2O$ | ⇋ | $CH_3COO^-$ | + | $H_3O^+$ |

Since both the forward and reverse reactions are occurring to some extent, acetic acid (ethanoic acid) which is donating a proton is considered a moderately weak acid, and acetate ions (ethanoate ions) which are accepting protons are considered moderately weak bases. The conjugate base of a moderately weak acid is a moderately weak base.

## Weak Acid - Strong Conjugate Base Pair

A weak Brønsted-Lowry acid has very little tendency to donate a proton. Weak Brønsted-Lowry acids include $H_2O_2$, $CH_3OH$ and $H_2O$. For example, methanol is a weak Brønsted-Lowry acid:

| Acid | + | Base | ⇋ | Conjugate base of $CH_3OH$ acid | + | Conjugate acid of $H_2O$ base |
|---|---|---|---|---|---|---|
| $CH_3OH$ | + | $H_2O$ | ⇋ | $CH_3O^-$ | + | $H_3O^+$ |

Since very little of the methanol dissociates to produce protons, the equilibrium position lies very far to the left:

| Acid | + | Base | ← | Conjugate base of $CH_3OH$ acid | + | Conjugate acid of $H_2O$ base |
|---|---|---|---|---|---|---|
| $CH_3OH$ | + | $H_2O$ | ← | $CH_3O^-$ | + | $H_3O^+$ |

The tendency for $CH_3O^-$ to gain a proton is much greater than the tendency for $CH_3OH$ to lose a proton, so $CH_3OH$ is a weak acid, but $CH_3O^-$ is a strong base. The conjugate base of a weak acid is strong base.

## Strong Base - Weak Conjugate Acid Pair

A strong Brønsted-Lowry base has a strong tendency to accept a proton. Strong Brønsted-Lowry bases include $OH^-$, $CH_3CO^-$ and $HO_2^-$.

For example, the hydroxide ion is a strong base:

| Base | + | Acid | ⇋ | Conjugate acid of $OH^-$ base | + | Conjugate base of acid $H_3O^+$ |
|---|---|---|---|---|---|---|
| $OH^-$ | + | $H_3O^+$ | ⇋ | $H_2O$ | + | $H_2O$ |

The equilibrium position for this reaction lies so very far to the right that we assume the reaction goes to completion:

| Base | + | Acid | → | Conjugate acid of $OH^-$ base | + | Conjugate base of acid $H_3O^+$ |
|---|---|---|---|---|---|---|
| $OH^-$ | + | $H_3O^+$ | → | $H_2O$ | + | $H_2O$ |

Since the water molecules formed as the conjugate acid of $OH^-$ have very little tendency to donate protons, they are only very weak acids. The conjugate acid of a strong base is only a weak acid.

## Moderately Weak Base - Moderately Weak Conjugate Acid Pair

A moderately weak Brønsted-Lowry base has only a slight tendency to accept a proton. Moderately weak Brønsted-Lowry bases include, $NH_3$, $CO_3^{2-}$, $HS^-$, $NO_2^-$ and $CH_3COO^-$.

For example, ammonia, $NH_3$, is a moderately weak base.

There is some tendency for $NH_3$ to accept a proton to form $NH_4^+$, but the $NH_4^+$ also has some tendency to donate a proton to form $NH_3$. The reaction forming $NH_4^+$ is in equilibrium with the reaction forming $NH_3$:

| Base | + | Acid | ⇋ | Conjugate acid of $NH_3$ base | + | Conjugate base of acid $H_2O$ |
|---|---|---|---|---|---|---|
| $NH_3$ | + | $H_2O$ | ⇋ | $NH_4^+$ | + | $OH^-$ |

Since both the forward and reverse reactions are occuring to some extent, it can be seen that the tendency for $NH_3$ to accept a proton is similar to the tendency for $NH_4^+$ to donate a proton, so that the $NH_3$ is considered to be a moderately weak base while the $NH_4^+$ is considered to be a moderately weak acid. The conjugate acid of a moderately weak base is a moderately weak acid.

**Weak Base - Strong Conjugate Acid Pair**

A weak Brønsted-Lowry base shows very little tendency to gain a proton. Weak Brønsted-Lowry bases include $H_2O$, $Cl^-$ and $NO_3^-$.

For example, the chloride ion is a weak Brønsted-Lowry base:

| Base | + | Acid | ⇋ | Conjugate acid of $Cl^-$ base | + | Conjugate base of acid $H_2O$ |
|---|---|---|---|---|---|---|
| $Cl^-$ | + | $H_2O$ | ⇋ | HCl | + | $OH^-$ |

Since $Cl^-$ displays almost no tendency to accept a proton, the equilibrium position lies very far to the left:

| Base | + | Acid | ← | Conjugate acid of $Cl^-$ base | + | Conjugate base of acid $H_2O$ |
|---|---|---|---|---|---|---|
| $Cl^-$ | + | $H_2O$ | ← | HCl | + | $OH^-$ |

This means that any HCl that might form, readily donates a proton to reform the $Cl^-$. $Cl^-$ has little tendency to gain a proton so it is a weak base, but its conjugate acid, HCl, has an enormous tendency to donate a proton it is a strong acid. The conjugate acid of a weak base is a strong acid.

To qualify as a Bronsted acid, a molecule must have:

- At least one H attached to an atom to the right of C in the Periodic Table
- Occasionally, we may need to write a Lewis structure to check this. For example, nitric acid traditionally is written $HNO_3$, suggesting that the H is attached to nitrogen. However

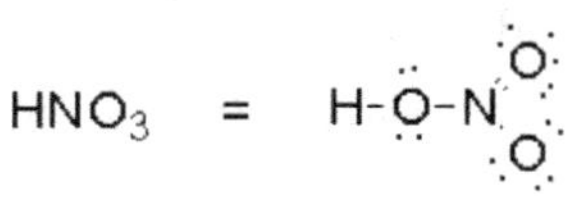


the Lewis structure shows the hydrogen actually is on oxygen.

A Bronsted base must have:

An atom with at least one unshared pair of electrons; again, sometimes one must write a Lewis structure to judge this.

The requirement of an unshared pair shows that the atom that binds the proton must lie to the right of B in the Periodic Table.

The strengths of acids then can be rationalized with a few basic rules:

- As the atom to which the acidic H is attached changes to the right in the Periodic Table, the acid gets stronger
- As the atom to which the acidic H is attached changes down a Group in the Periodic Table, the acid gets stronger
- When the oxygen the H is attached to is part of a polyatomic group (For example, $H_2SO_4$), the greater the number of oxygens in the group, the stronger the acid
- Reverse these rules for basicity, since, as we have noted, strong acid have weak conjugate bases, and *vice versa*.

The basis for these rules is that we want the **bond** to the acidic H to be polar:

$$\underset{\delta+}{\mathrm{H}} - \underset{\delta-}{\mathrm{A}} \quad \text{A Polar bond}$$

or weak having low bond dissociation energy. Polarity controls the trend across the Periodic Table, bond dissociation energy controls it down a Group.

## 4.4 Acid-base Properties

Acids and bases are relative terms. A molecule cannot be called an acid in exclusion of all else. For example, the gaseous hydrogen chloride molecule HCl(g) is a very stable molecule and does not act as acid or base unless something else is introduced to the system. If water is introduced to the system, the HCl readily reacts with the water to form $H_3O^+$(aq) and $Cl^-$(aq). In other words, in the process of dissolving into the water, the HCl molecule gives up a proton to become $Cl^-$(aq) and a water molecule accepts the proton and becomes $H_3O^+$(aq).

Bases are the chemical opposite of acids. Acids are defined as compounds that donate a hydrogen ion ($H^+$) to another compound (base). Traditionally, an acid was any chemical compound that, when dissolved in water, gives a solution with a hydrogen ion activity greater than in pure water, i.e. a pH less than 7.0. Correspondingly, a base was any compound that, when dissolved in water, gives a solution with a hydrogen ion activity lower than that of pure water, i.e. a pH higher than 7.0 at standard conditions.

A soluble base is also called an alkali. A reaction between an acid and a base is called neutralization and this neutralization results in production of water and a salt. Volatile liquids (acids) when mixed with specific substances turn into salts.

These substances form a concrete base and hence the name base was derived. Acids in general are $H^+$ donors and Bases are $H^+$ acceptors.

Bases have a slippery feel on fingers and taste bitter. They change litmus paper blue. Acids taste sour and create a stinging feeling on the mucous membranes. They change litmus paper red. They can react with bases to produce salts and water. They both conduct electricity depending on the dissociation of ions. Acids have a pH lesser than 7.0 and the lower it is, the stronger the acid becomes. Bases have a pH between 7 and 14. Higher the pH value, stronger will be the base. A pH level of 7 is a neutral substance which is water.

### 4.4.1 Acid-Base Properties of Oxides

Most elements of the periodic table form at least one oxide. The oxides of the representative metals are ionic in character. They dissolve in water to form basic solutions. The oxides of the nonmetals are generally covalent in character. Many of them react with water to yield acidic solutions. Many nonmetals form more than one oxide. In general, the acidity of the acid formed from the oxide increases with increasing oxidation state of the nonmetal.

For example, sulfur forms two common oxides: sulfur dioxide, in which sulfur is in the +4 oxidation state and sulfur trioxide in which it is in the +6 oxidation state. Sulfur dioxide reacts with water to form sulfurous acid, $H_2SO_3$, a weak acid. Similarly, $SO_3$ reacts with water under appropriate conditions to yield sulfuric acid, a strong acid. In summary, the oxides of the strongly metallic elements form basic solutions, while the oxides of the nonmetals generally for acidic solutions. The oxides of many other elements may exhibit either acidic or basic character and are said to be amphoteric.

Compounds of any element with water are called **oxides**.

**Acidic Oxides**

They are oxides which combine with bases and form salt. $SO_2$, $SO_3$, $CO_2$, $N_2O_5$ are example of acid oxides.

$$SO_3 + 2\ NaOH \rightarrow Na_2SO_4 + H_2O$$

Acid Oxide Base Salt Water

$$CO_2 + 2\ NaOH \rightarrow Na_2CO_3 + H_2O$$

Acid Oxide Base Salt Water

Acid oxides combine with water and form acids.

$$SO_2 + H_2O \rightarrow H_2SO_3$$

$$SO_3 + H_2O \rightarrow H_2SO_4$$

$$CO_2 + H_2O \rightarrow H_2CO_3$$

**Basic Oxides**

They combine with acids and form salt. Metal oxides show this property like; $Na_2O$, CaO.

$$Na_2O + 2HCl \rightarrow 2NaCl + H_2O$$

Basic Oxide Acid Salt Water

$$CaO + 2HCl \rightarrow CaCl_2 + H_2O$$

Basic Oxide Acid Salt Water

Basic oxides combine with water and form bases.

$$Na_2O + H_2O \rightarrow 2NaOH$$

$$BaO + H_2O \rightarrow Ba(OH)_2$$

**Neutral Oxides**

They are do not react with acids and bases. Neutral oxides do not react with water and form acid or base. NO, $N_2O$ and CO are some examples of neutral oxides.

**Amphoteric Oxides**

These oxides react with acids and bases and form salt. ZnO and $Al_2O_3$ are examples of amphoteric oxides. Examples of these reactions are given below;

$$ZnO + 2HCl \rightarrow ZnCl_2 + H_2O$$

$$ZnO + 2NaOH \rightarrow Na_2ZnO_2 + H_2O$$

$$Al_2O_3 + 6HCl \rightarrow 2AlCl_3 + 3H_2O$$

$$Al_2O_3 + 6NaOH \rightarrow 2Na_3AlO_3 + 3H_2O$$

**Peroxides**

Compounds including $(O_2)^{-2}$ in their structure are called peroxides. Examples of peroxides are given below;

$H_2O_2$: Hydrogen Peroxide

$Na_2O_2$: Sodium Peroxide

**Example**

Which ones of the following statements are true for water solution of sulfuric acid $H_2SO_4$?

i) It turns color of litmus to red.

ii) It conducts electric current.

iii) When it reacts with Mg, $H_2$ gas is formed.

iv) It does neutralization reaction with water solution of $NH_3$.

**Solution**

Since it is acid, it turns blue litmus to red I is true.

All of the acidic water solutions conduct electric current so II is true.

Some of the metals react with acids and $H_2$ gas is formed, Mg is one of that metals III is also true.

$NH_3$ is base and $H_2SO_4$ is acid, when they combine neutralization reaction occurs. IV is also true.

## Reaction of Amphoteric Hydroxides in Acidic Solution

One of the most common and familiar examples of an amphoteric hydroxide is aluminum hydroxide, $Al(OH)_3$. From our solubility rules, we know that $Al(OH)_3$ is largely insoluble in neutral water; however, in a strongly acidic solution, the situation changes. For example, consider the reaction of $Al(OH)_3$ with HCl:

$$HCl(aq) + Al(OH)_3(aq) \rightarrow AlCl_3(aq) + 3H_2O(l)$$

This is a classic acid-base neutralization reaction: the HCl completely protonates all three hydroxides per mole of $Al(OH)_3$, yielding pure water and the salt $AlCl_3$.

Reaction of Amphoteric Hydroxides in Basic Solution

We will now consider aluminum hydroxide's reaction in a strongly basic Solution

$$Al(OH)_3(aq) + OH^-(aq) \rightarrow Al(OH)^-_4(aq)$$

Here, aluminum hydroxide takes hydroxide ion out of solution, thereby acting as a Lewis acid. In Lewis structure for $Al(OH)_3$ the central aluminum atom is electron deficient as it forms only three bonds while the octet rule is not fulfilled, so $Al^{3+}$ is quite happy to accept a pair of electrons and form another bond, giving the right conditions. In a basic solution, it forms a bond with an $OH^-$ ion, pulling it out of solution and lowering the solution's pH.

## 4.4.2 Acid-Base Properties of Salts

The reaction between an acid and a base results in a salt and water results four possible cases:

### Salt of Strong Acid and Strong Base

Consider an example, hydrochloric acid and sodium hydroxide. If they are added together in aqueous solution such that they exactly neutralize each other, the result will be a NaCl solution in water.

$$HCl + NaOH \rightarrow NaCl + HH_2O$$

In this case, the solution is the result of the neutralization of a strong acid with a strong base and the pH is exactly 7. In other words, the salt NaCl is neither acidic nor basic. The $Na^+$ is a very weak acid (the conjugate acid of a strong base, NaOH), in other words, it is not an acid. The $Cl^-$ ion is a very weak base (the conjugate base of a strong acid, HCl), and is therefore not a base. So NaCl is a salt made up of a cation that is not an acid and an anion that is not a base. It is neutral. Such is not the case if the original acid and base are not both strong..

**Salt of Strong Acid and Weak Base**

Take, for example, the reaction between Hydrochloric acid and ammonia, a weak base.

$$HCl + NH_3 \rightarrow NH_4Cl$$

Here, we still have the anion $Cl^-$ that we know to be not a base. the cation, $NH_4^+$, however is a weak acid. It's the conjugate acid of the weak base $NH_3$, since the conjugate of a weak base is always a weak acid. Thus, the salt ammonium chloride is made up of a cation ammonium which is a weak acid and an anion $Cl^-$ that is not a base. Thus, the solution of ammonium chloride is acidic.

**Salt of Weak Acid and Strong Base**

In the solution of sodium acetate, $NaCH_3COO$, cation is $Na^+$ which is neutral as discussed, so the anion is acetate $CH_3COO^-$, which is conjugate base of weak acid acetic acid $CH_3COOH$. As acetic acid is weak, acetate is also weak, hence sodium acetate is made up of neutral cation and a basic anion, so the solution is basic.

**Salt of Weak Acid and Weak Base**

In ammonium ion $NH_4^+$ is acidic and the acetate ion $CH_3COO^-$ is basic, so to have an answer to the question, compare relative strengths of acidic cation and basic anion. In other words, if $K_a > K_b$ the solution will be acidic. If $K_a < K_b$ then the solution is basic. It turns out, that in this case, the two happen to be almost equal so the salt sodium acetate is almost neutral.

## 4.4.3 Acid-Base Properties of Organic Compounds

Acid and base properties of organic compounds are much the same as acid and base properties of inorganic compounds. There are both strong and weak inorganic acids and bases, however, while organic acids and bases are universally weak acids and weak bases. Thus the inorganic acids and bases don't exhibit the acid and base properties as 'strongly' as the strong mineral acids and inorganic bases.

Properties of acids include:

- They have a pH less than 7.
- They have a sour taste.
- They produce hydrogen ions when dissolved in water.

- They are corrosive to human tissue (i.e. they 'burn' your skin).
- They react with bases to form a salt and water.

Common properties of bases (either organic bases or inorganic bases) include:

- They have a pH more than 7.
- They have a 'soapy' feel.
- They have a bitter taste.
- They are corrosive to human tissue (i.e. they 'burn' your skin).
- They react with acids to form a salt and water.

**Organic Acids**

Organic acids are, in general, weak acids. That is, they do not ionize nearly completely in weak aqueous solution. Some inorganic acids are strong acids. That is, they ionize nearly completely in weak aqueous solution. Some common strong mineral acids are hydrochloric acid, sulfuric acid, and nitric acid. The acid properties listed above are 'stronger' or more noticeable for the strong mineral acids than for the weak organic acids. Strong acids will have a pH much lower than 7, will produce many hydrogen ions, and will be very corrosive to the skin. Weak organic acids, on the other hand, will have a pH closer to 7 than a strong acid; will produce some hydrogen ions when dissolved in water, but not as many as a strong acid; and will not be as corrosive to the skin as a strong acid.

Most common organic acids contain a carboxyl group, usually written as - COOH. The structure of the carboxyl group is shown in the diagram at the left. The hydrogen-oxygen bond in the OH is the weakest bond in the carboxyl group, so Hydrogen ions are released from carboxyl containing organic acids when they are dissolved in water. One of the simpler organic acid is acetic acid (vinegar) with the formula: $CH_3COOH$ (or $C_2H_4O_2$). An expanded version of the formula, showing the carboxyl group is shown in the diagram at the right.

Some other common organic acids are:

- Propionic acid - $CH_3CH_2COOH$ (or $C_3H_6O_2$)
- Lactic acid - $CH_3CH(OH)COOH$ (or $C_3H_6O_3$)
- Formic acid - $HCOOH$ (or $CH_2O_2$)
- Oxalic acid - $COOHCOOH$ (or $H_2C_2O_4$)
- Citric acid - $COOHCH_2C(OH)COOHCH_2COOH$ (or $C_6H_8O_7$)

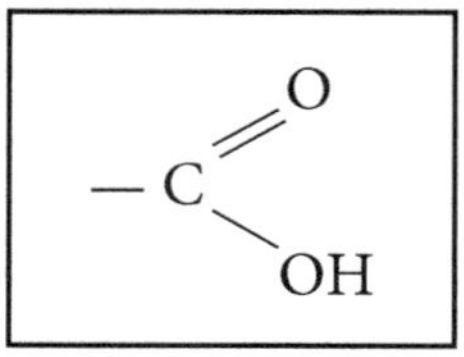


Carboxyl Group

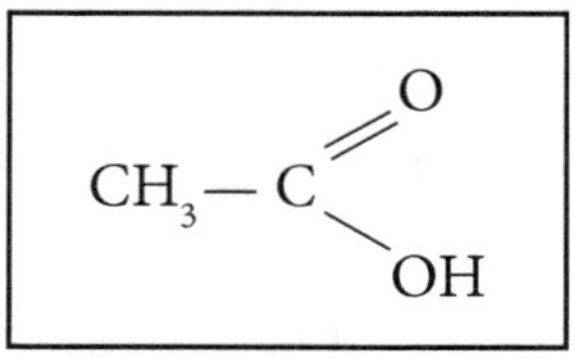


Acetic Acid

Organic Acids.

The formulas for each of these organic acids are shown first in a format that emphasizes the functional groups (including COOH at least once in each of them) and then in the more compact, less descriptive format that you will often see.

**Organic Bases**

Organic bases typically do not ionize nearly completely in weak aqueous solution, so they are classified as weak bases. This is in contrast to strong bases like sodium hydroxide or potassium hydroxide, which do ionize nearly completely in weak aqueous solution. Just as with the weak and strong acids, the weak organic bases don't exhibit the properties of bases as 'strongly' as strong bases.

Organic bases exhibit their basic behavior by having an attraction for hydrogen ions, usually through one or more nitrogen atoms in the molecule.

Some common organic bases are:

- Methyl amine - $CH_3NH_2$ (or $CH_5N$)
- Pyridine - $C_5H_5N$
- Imidazole - $C_3H_4N_2$
- Glycine - $C_2H_5NO_2$

# 5 Monoprotic and Polyprotic Acids and Bases

## 5.1 Introduction

Acids are defined in several ways by various scientists. Arrhenius defines an acid as a substance that donates $H_3O^+$ ions in the solution. Bronsted- Lowry defines a base as a substance that can accept a proton. Lewis acid definition is far common than the above two. According to it, any electron pair donator is a base. According to the Arrhenius or Bronsted-Lowry definition, a compound should have a hydrogen and the ability to donate it as a proton to be an acid. However, according to Lewis, there can be molecules, which don't possess hydrogen, but can act as an acid. For example, $BCl_3$ is a Lewis acid, because it can accept an electron pair. An alcohol can be a Bronsted-Lowry acid because it can donate a proton but, according to Lewis, it will be a base.

Regardless of the above definitions, we normally identify an acid as a proton donor. Acids have a sour taste. Lime juice, vinegar are two acids we come across at our homes. They react with bases producing water, and they also react with metals to form $H_2$, thus increase metal corrosion rate. Acids can be categorized into two, based on their ability to dissociate and produce protons. Strong acids like HCl, $HNO_3$ are completely ionized in a solution to give protons. Weak acids like $CH_3COOH$ partially dissociate and give fewer amounts of protons. $K_a$ is the acid dissociation constant. It gives an indication of the ability to lose a proton of a weak acid. To check whether a substance is an acid or not we can use several indicators like litmus paper or pH paper. In the pH scale, from 1-6 acids are represented. An acid with pH 1 is said to be very strong, and as the pH value increases, acidity is decreased. Moreover, acids turn blue litmus to red.

## 5.2 Features of Monoprotic Acids and Base

As per Brønsted and Lowry, acid is a proton donor and base is proton acceptor. This idea of proton donor and proton acceptor is important in understanding monoprotic and polyprotic acids and bases because monoprotic corresponds to the transfer of one proton and polyprotic refers to the transfer of more than one proton. Therefore, a monoprotic acid is an acid that can donate only one proton, while polyprotic acid can donate more than one proton. Similarly, a monoprotic base can only accept one proton, while a polyprotic base can accept more than one proton.

When one molecule of acid dissociates in an aqueous solution, if it gives a single proton, then that acid is said to be a monoprotic acid. HCl and nitric acid ($HNO_3$) are some examples for monoprotic mineral acids. Following is the dissociation for HCl in the aqueous medium to give out one proton.

$$HCl \rightarrow H^+ + Cl^-$$

Other than the mineral acid, there can be monoprotic organic acids too. Usually when there is one carboxylic group, that acid is monoprotic. For example, acetic acid, benzoic acid, and a simple amino acid like glycine are monoprotic.

A monoprotic acid is an acid that donates only one proton or hydrogen atom per molecule to an aqueous solution. This is in contrast to acids capable of donating more than one proton or hydrogen, which are called polyprotic acids. Polyprotic acids may be further categorized according to how many protons they can donate (diprotic = 2, triprotic = 3, etc.).

The electrical charge of a monoprotic acid is one level higher before it gives away its proton.

Any acid that one contains one hydrogen atom in its formula is monoprotic. However, some acids that contain more than one hydrogen atom are monoprotic. Because only one hydrogen is released, the pH calculation for a monoprotic acid is straightforward.

The Bronsted Lowry theory states that acids will be proton donors while bases will be proton acceptors. Thus, a monoprotic acid is an acid that can be classified as a proton donor. However, there is one caveat to this definition. A monoprotic acid can only donate one acidic proton. This should make perfect sense given that the prefix mono in monoprotic acid means 'one.'

Acids are proton donors while Bases are proton acceptors

$$HCl(g) + NH_3(g) \rightarrow NH_4^+Cl^-(s)$$

acid base salt

$$CH_3OOH + CH_3NH_2 \leftrightarrow CH_3OO^- + CH_3NH_3^+$$

Acetic acid methyl amine acetate ion methyl ammonium ion

acid base conjugate base conjugate acid

For aqueous solutions,

Acids react with water to yield $H_3O^+$ while bases react with water to yield $OH^-$ acid

| $CH_3OOH + H_2O$ | | $\leftrightarrow$ | $CH_3OO^- +$ | $H_3O^+$ |
|---|---|---|---|---|
| Acetic acid | water | | acetate ion | hydronium ion |
| acid | base | | conjugate base | conjugate acid |

conjugate pairs

$CH_3OOH$

$CH_3OO^-$

base

| $CH_3NH_2$ | + | $H_2O$ | $\leftrightarrow$ | $CH_3NH_3^+$ | + | $OH^-$ |
|---|---|---|---|---|---|---|
| methyl amine | | water | | methyl ammonium ion | | hydroxide ion |
| base | | acid | | conjugate acid | | conjugate base |

conjugate pairs

$CH_3NH_2$

$CH_3NH_3^+$

### 5.2.1 Autoprotolysis of Water

$$H_2O + H_2O \leftrightarrow H_3O^+ + OH^-$$

Equilibrium constant is Kw

$$Kw = [H_3O^+][OH^-] = 1.01*10^{-14} \text{ @ } 25 \text{ C}$$

**For pure water**

$$[H_3O^+] = [OH^-] = 1.00*10^{-7} \text{ M}$$

**pH – a measure of the acidity of a solution**

$$pH = -\log [H_3O^+] \text{ or } -\log A_{H+}$$

below 7 is acidic

above 7 is neutral

### 5.2.2 Strong Acids and Bases – Complete Dissociation in Water

**Example**

What is the pH of a 0.2 M HCl Solution?

**Solution**

HCl is a strong acid

$$HCl + H_2O \rightarrow Cl^- + H_3O^+$$

The equilibrium constant for this reaction is large. There is virtually no HCl species remaining in solution.

Therefore,

$[H_3O^+] = 0.2$ M and pH = –log (0.2) = 0.7

0.02 M solution → 1.7

As per sig. fig. rule, number of significant figures in concentration is equal to number of significant figures after decimal point in the pH.

**Example**

What is the pH of a 0.020 M $Ca(OH)_2$ solution?

**Solution**

$Ca(OH)_2$ is a strong base.

$$Ca(OH)_2 \rightarrow Ca^{2+} + 2OH^-$$

The equilibrium constant for this reaction is large. There is virtually no $Ca(OH)_2(s)$ remain, so

$[OH^-] = 0.040$ M, $[H_3O^+]$

$= 2.5*10^{-13}$ M

pH = –log (0.040)

= 12.60

As per sig. fig. rule, number of significant figures in concentration is equal to number of significant figures after the decimal point in the pH. However, in examples, there exist assumption that has implicitly been made is that the $[H^+]$ from the strong acid (or $[OH^-]$ from the strong base) is the [H+] (or the $[OH^-]$) from water hydrolysis.

In the strong acid example,

$[H^+]_{sa} = 0.1$ M

$Kw = [OH^-][H_3O^+] = 1.01*10^{-14}$

$[OH^-] = 1.01*10^{-13}$ M

$[H_3O^+]_{water} = [H^+]_{water} = 1.01*10^{-13}$ M

**Example**

Calculate the pH of a $5.0*10^{-8}$ M $HClO_4$ solution

**Solution**

$HClO_4$ is a strong acid, and thus completely dissociates to give $H_3O^+$ and $ClO_4^-$ ions.

Sources of $H_3O^+$

$$HClO_4 + H_2O \rightarrow H_3O^+ + ClO_4^-$$

$$2H_2O \leftrightarrow H_3O^+ + OH^-$$

pertinent equations;

$$[H_3O^+] = [H_3O^+]_{HClO4} + [H_3O^+]_{H2O}$$

$$[OH^-] = [H_3O^+]_{H2O}$$

$$1.01*10^{-14} = Kw = [H_3O^+]\ [OH^-]$$

$$[H_3O^+]_{HClO4} = 5.0*10^{-8}\ M$$

let $x = [H_3O^+]_{H2O}$ substitutions;

$$Kw = x(5.0*10^{-8} + x) = 1.01*10^{-14}$$

Assumption 1: Is $[H_3O^+]_{H2O} <<< [H_3O^+]_{HA}$

If at first we make the assumption, we soon see that:

$$[H_3O^+]_{HA} = 5.0*10^{-8}\ M$$

$$[H_3O^+]_{H2O} = 2.0*10^{-7}\ M$$

Assumption is not valid

Considering original expression,

$$x(5.0*10^{-8} + x) = 1.01*10^{-14}$$

This is a quadratic, so applying successive approximations,

$$x = 7.86*10^{-8}\ M$$

$$[H_3O^+] = 5.0*10^{-8} + x = 1.286*10^{-7}\ M,$$

$$pH = -\log(1.286*10^{-7}) = 6.89$$

The pH of a dilute strong acid ought to be slightly acidic.

### 5.2.3 Weak Acids and Bases

**Treatment of Weak acids – HA**

$$HA + H_2O \leftrightarrow A^- + H_3O^+$$

Equilibrium constant is $K_a$

$$K_a = [A^-][H_3O^+]/[HA]$$

The autoprotolysis reaction must also be satisfied

$$2H_2O \leftrightarrow OH^- + H_3O^+$$

$Kw = [OH^-][H_3O^+] = 1.01*10^{-14}$

$K_a = [A^-]([H_3O^+]_{HA} + [H_3O^+]_{H2O})/([HA]_0 - [A^-])$

Define: $[H_3O^+]_{HA}$

$[H_3O^+]_{H2O}$

$[HA]_0$

**Assumptions to be made**

$[H_3O^+]_{H2O} <<< [H_3O^+]_{HA}$, therefore HA is the only source of $H_3O^+$, and $\mathbf{[A^-] = [H_3O^+]}$

If Ka is small, $[A^-] <<< [HA]$, and $\mathbf{[HA] = [HA]_0}$

**Example**

Calculate the pH of a 0.10 M solution of acetic acid.

$K_a = 1.75*10^{-5}$

$CH_3OOH + H_2O \leftrightarrow CH_3OO^- + H_3O^+$

**Solution**

$K_a = 1.75*10^{-5} = [CH_3OO^-]\ [H_3O^+] / [CH_3OOH]$

Let $x = [CH_3OO^-]$

$K_a = x(x + [H_3O^+]_{H2O}) / (0.10 - x)$

Making assumptions:

$[H_3O^+]_{H2O} <<< [H_3O^+]_{HA}$

$[HA] = [HA]_0$

The Ka expression reduces to

$K_a = x^2 / 0.10 = 1.75*10^{-5}$

$x = 0.001322876\ M = [CH_3OO^-] = [H_3O^+]$

pH = -log(0.001322876) = 2.88 (2 sig fig after the decimal pt.)

Considering Assumption 1:

Is $[H_3O^+]_{H2O} <<< [H_3O^+]_{HA}$?

$[H_3O^+] = 0.00132$ M, so $[OH^-] = Kw/[H_3O^+] = 7.65*10^{-12}$ M

The only source of the $[OH^-]$ is from autoprotolysis of water, therefore,

$[H_3O^+]_{H2O} = [OH^-]$

$= 7.65*10^{-12}$ M

Hence, assumption 1 is valid

Considering Assumption 2:

Is $[HA] >> [A^-]$?

If $[A^-]/[HA]_0 < 0.05$, then assumption is valid.

In other words, extent of dissociation is less than 5 %

$$[A^-]/[HA]_0 = (.00132/0.1)$$

$$= 0.0132, \text{ assumption is valid}$$

### 5.2.4 Treatment of Weak bases, B

$$B + H_2O \leftrightarrow BH^+ + OH^-$$

Equilibrium constant is Kb

$$K_b = [BH^+][OH^-]/[B]$$

The autoprotolysis reaction must also be satisfied

$$2H_2O \leftrightarrow OH^- + H_3O^+$$

$$Kw = [OH^-][H_3O^+] = 1.01*10^{-14}$$

$$K_b = [BH^-]([OH^-]_B + [OH^-]_{H2O})/([B]_0 - [BH^+])$$

Define: $[OH^-]_{HA}$

$[OH^-]_{H2O}$

$[B]_0$

Assumptions that can be made:

$[OH^-]_{H2O} <<< [OH^+]_B$, therefore B is the only source of $OH^-$, and $[BH^+] = [OH^-]$

If $K_b$ is small, $[BH^+] <<< [B]$, and $[B] = [B]_0$

**Example**

Calculate the pH of a 0.10 M solution of sodium acetate.

**Solution**

Sodium acetate is a soluble salt

$$NaAc \rightarrow Na^+ + Ac^-$$

$$CH_3OO^- + H_2O \leftrightarrow CH_3OOH + CH^-$$

$K_b$(acetic acid) is the equilibrium constant for this reaction

$K_b$(acetic acid) = $Kw/K_a$(acetic acid) = $1.01*10^{-14}/1.75*10^{-5} = 5.77*10^{-10}$

$K_b = 5.77*10^{-10} = [CH_3OOH]\ [OH^-] / [CH_3OO^-]$

Let $x = [CH_3OOH]$

$K_b = x(x + [OH^-]_{H2O}) / (0.10 - x)$

Making assumptions:

1. $[OH^-]_{H2O} <<< [OH^-]_{A-}$
2. $[A^-] = [A^-]_0$

The $K_b$ expression reduces to

$K_b = x^2 / 0.10 = 5.77*10^{-10}$

$x = 7.596991886*10^{-6}\ M = [CH_3OOH] = [OH^-]$

$[H_3O^+] = Kw/[OH-]$

$= 1.01*10^{-14}/7.596991886*10^{-6}$

$= 1.3294735*10^{-9}\ M$

$pH = -\log(1.3294735*10^{-9})$

$= 8.88$ (2 sig fig after the decimal pt.)

Considering Assumption 1:

Is $[OH^-]_{H2O} <<< [OH^-]_{HA}$?

$[OH^-] = 7.60*10^{-6}\ M$

therefore, $[OH^-]_{H2O} = [H_3O^+] = 1.3*10^{-9}\ M$

If $[OH^-]_{H2O} / [OH^-] < 0.05$, then it is observed that assumption is valid

$[OH^-]_{H2O} / [OH^-] = 1.3*10^{-9}/7.60*10^{-6} = 0.0002$

Assumption 1 is valid

Considering Assumption 2:

Is $[B] >> [BH^+]$?

If $[BH^+]/[B]_0 < 0.05$, we will call the assumption valid. In other words; the extent of dissociation is less than 5 %

$[BH^+]/[B]_0 = (7.6*10^{-6}/0.1) = 7.6*10^{-5}$, assumption is valid

**Example**

What is the pH of a 5.0 mM solution of ammonia?

**Solution**

Ammonia is a weak base

$K_b = Kw/Ka_{(NH^{4+})} = 1.01*10^{-14}/5.70*10^{-10} = 1.77*10^{-5}$

$CH_3OO^- + H_2O \leftrightarrow CH_3OOH + OH^-$

$K_b = 1.77*10^{-5} = [NH_4^+] [OH^-] / [NH_3]$

Let $x = [NH_4^+]$

$K_b = x(x + [OH^-]_{H_2O}) / (0.005 - x)$

Now making assumptions:

1. $[OH^-]_{H2O} <<< [OH^-]_{NH3}$
2. $[NH_3] = [NH_3]_0$

The $K_b$ expression reduces to

$K_b = x^2 / 0.0050 = 1.77*10^{-5}$

$x = 2.97489*10^{-4}$ M $= [CH_3OOH] = [OH^-]$

$[H_3O^+] = Kw/[OH^-] = 1.01*10^{-14}/2.97489*10^{-4} = 3.3950779*10^{-11}$ M

pH = $-\log(3.3950779*10^{-11}) = 10.47$ (2 sig fig after the decimal pt.)

Considering Assumption 1:

Is $[OH^-]_{H2O} <<< [OH^-]_{HA}$?

$[OH^-] = 2.97*10^{-4}$ M

Hence, $[OH^-]_{H_2O} = [H_3O^+]$

$= 3.40*10^{-11}$ M

If $[OH^-]_{H2O} / [OH^-] < 0.05$, we will call the assumption valid

$[OH^-]_{H2O} / [OH^-] = 3.40*10^{-11}/2.970*10^{-4} <<<< .05$

So assumption 1 is valid

Considering Assumption 2:

Is $[B] >> [BH^+]$?

If $[BH^+]/[B]_0 < 0.05$, we will call the assumption valid. In other words; the extent of dissociation is less than 5 %

$[BH^+]/[B]_0 = (2.970*10^{-4} /0.005) = 0.06$,

Hence assumption is not valid.

Now writing expression without the assumption:

$K_b = x^2 / (0.0050\text{-}x) = 1.77*10^{-5}$

This is a quadradic

$x^2 + 1.77*10^{-5}x - (0.005)(1.77*10^{-5})$

On solving:

$x = 2.8877 *10^{-4}$ M *instead of (2.970*10⁻⁴)*

$[H_3O^+] = 1.01*10^{-14} / 2.8877 *10^{-4}$

$= 3.4976*10^{-11}$

pH = 10.46

**Example**

Calculate the pH of a 0.010 M solution of chlorous acid.

**Solution**

$K_a = 1.12*10^{-2}$

$HOClO + H_2O \leftrightarrow OClO^- + H_3O^+$

$K_a = 1.12*10^{-2} = [OClO^-] [H_3O^+] / [HOClO]$

Let $x = [CH_3OO^-]$

$K_a = x(x + [H_3O^+]_{H2O}) / (0.010 - x)$

Making assumptions:

1. $[H_3O^+]_{H2O} <<< [H_3O^+]_{HA}$
2. $[HA] = [HA]_0$

The Ka expression reduces to

$K_a = x^2 / 0.010 = 1.12*10^{-2}$

$x = 0.0105838$ M $= [CH_3OO^-] = [H_3O^+]$

Considering Assumption 1:

Is $[H_3O^+]_{H_2O} <<< [H_3O^+]_{HA}$?

$[H_3O^+] = 0.0106$ M, so $[OH^-] = Kw/[H_3O^+] = 9*10^{-13}$ M

The only source of the $[OH^-]$ is from autoprotolysis of water, so,

$[H_3O^+]_{H_2O} = [OH^-]$

$= 7.65*10^{-12}$ M

Hence assumption 1 is valid

Considering Assumption 2:

Is $[HA] >> [A^-]$?

Now solving expression for x

$$1.12*10^{-2} = x(x + [H_3O^+]_{H2O}) / (0.010 - x)$$

$$x = 0.006181 \text{ M}$$

$$pH = 2.21$$

### 5.2.5 Buffers

A buffer is a solution that is resistant to pH changes. It usually consist of a mixture of an acid and its conjugate base in roughly equal proportions. Let's look at weak acid, HA ($K_a = 1.00*10^{-4}$)

$$HA + H_2O \leftrightarrow A^- + H_3O^+$$

$$K_a = [A^-] [H_3O^+] / [HA] = [H_3O^+] \{[A^-] / [HA] \}$$

Take the log of both sides

$$\text{Log Ka} = \log [H_3O^+] + \log \{[A^-] / [HA]\}$$

$$-pKa = -pH + \log \{[A^-] / [HA]\}$$

$$pH = pK_a + \log \{[A^-] / [HA]\} \text{ Henderson-Hasselbach equation}$$

It is observed that for solution containing an equal mixture of an acid and its conjugate base, pH = pKa, which is independent of pH. As [HA] and $[A^-]$ are components of the same solution, so on rewriting Henderson-Hasselbach equation:

$$pH = pK_a + \log (\text{mol } A^-] / \text{mol HA})$$

Add 1.0 ml of concentrated HCl to 1.0 L of water and find the pH.

Concentrated HCl is 12.1 M

1 ml of concentrated HCl = 12.1 mmol

$$[H_3O^+] = (12.1 \text{ mmol})/1001 \text{ ml} = 0.012 \text{ M}$$

$$pH = -\log(.012) = 1.92$$

Now add 1.0 ml of concentrated HCl to 1.0 L solution containing 0.50 mol of acetic acid (HA) and 0.50 mol of sodium acetate (NaA).

$$K_{a(\text{acetic acid})} = 1.75*10\text{-}5$$

$$pKa = 4.757$$

pH before the addition of the HCl

$$pH = 4.757 + \log(\text{mol } A^- / \text{mol HA})$$

= 4.757 + log (0.50/0.50)

= 4.76

Now pH after the addition of HCl, it shows that HCl will react rapidly and completely with $A^-$ (it is a strong acid)

$HCl + A^- \rightarrow Cl^- + HA$

Considering initial conditions:

HCl: 12.1 mmol

$A^-$: 500 mmol

HA: 500 mmol

Considering final conditions:

HCl: 0.00 mmol

$A^-$: (500-12.1) = 487.9 mmol

HA: (500+12.1) = 512.1 mmol

Now pH = 4.757 + log(mol $A^-$ / mol HA)

= 4.757 + log (487.9/512.1)

= 4.74

**Example**

Add 1.0 ml of concentrated HCl to 1.0 L solution containing 0.050 mol of acetic acid (HA) and 0.050 mol of sodium acetate (NaA) and find the pH?

**Solution**

Initial conditions:

HCl: 12.1 mmol

$A^-$: 50 mmol

HA: 50 mmol

Final conditions:

HCl: 0.00 mmol

A-: (50-12.1) = 37.9 mmol

HA: (50+12.1) = 62.1 mmol

Now the pH = 4.757 + log(mol $A^-$ / mol HA)

= 4.757 + log (37.9/62.1)

= 4.54

**Example**

Add 1.0 ml of concentrated HCl to 1.0 L solution containing 0.0050 mol of acetic acid (HA) and 0.0050 mol of sodium acetate (NaA) and find pH?

**Solution**

Initial conditions:

HCl: 12.1 mmol

$A^-$: 5.0 mmol

HA: 5.0 mmol

Final conditions:

HCl: 7.1 mmol

A-: (5.0-5.0) = 0 mmol

HA: (5.0+5.0) = 10 mmol

pH = -log(.0071)

= 2.2

the capacity of the buffer solution was exceeded

It is observed that buffer capacity (b)

= dCb/dpH

= –dCa/dpH

Buffer capacity depends on

- ratio of $A^-$/HA, closer to 1, the greater the buff. Capacity
- the absolute concentration of $A^-$ and HA

**Example**

Add 1.0 g of concentrated NaOH to 1.0 L solution containing 0.050 mol of acetic acid (HA) and 0.050 mol of sodium acetate (NaA). Find the pH?

**Solution**

$$NaOH + HA \rightarrow Na^+ + H_2O + A^-$$

In this, NaOH will react rapidly and completely with HA (it is a strong base), hence finding initial and final condition.

Initial conditions:

NaOH: (1.0 g)/(40.0 g/mol) = 25 mmol

$A^-$: 50 mmol

HA: 50 mmol

Final conditions:

NaOH: 0.00 mmol

A-: (50+25) = 75 mmol

HA: (50-25) = 25 mmol

Now pH = 4.757 + log(mol $A^-$ / mol HA)

= 4.757 + log (75/25)

= 5.23

**Example**

What is the pH of a solution prepared by mixing 0.0100 mol HA (with Ka = $1.00*10^{-2}$) and 0.0100 mol of $A^-$ and diluting with water to 1.00 L?

**Solution**

Because HA is a moderately strong acid, the following reaction can occur to an appreciable extent.

$$HA + H_2O \leftrightarrow A^- + H_3O^+$$

$$Ka = [H_3O^+][A^-]/[HA]$$

$$pH = pK_a + \log \{[A^-] / [HA]\}$$

$$[A^-] = F_{A-} + [H_3O^+]$$

$$[HA] = F_{HA} - [H_3O^+]$$

$$F_{A-} = F_{HA} = 0.0100 \text{ M}$$

Let $[H_3O^+] = x$

$$K_a = x(.0100+x)/(.0100-x)$$

$$1.00*10^{-2} = x(.0100+x)/(.0100-x)$$

With successive approximations → x = 0.00414 M

$$[A^-] = F_{A-} + [H_3O^+]$$

$$= 0.0100 + 0.00414$$

= 0.01414 M

Now [HA] = $F_{HA}$ - [$H_3O^+$]

= 0.0100 – 0.00414

= 0.00586 M

Hence pH = $pK_a$ + log {[$A^-$] / [HA]}

= 2.000 + log(0.01414/0.00586)

= 2.383

## 5.3 Polyprotic Acids Characteristics

Polyprotic acids contain more than one hydrogen atoms, which can be donated as protons when they are dissolved in an aqueous medium. Specifically, if they are donating two protons, we call them as diprotic and, if giving three protons, triprotic, etc.

Hydrogen sulfide ($H_2S$) and $H_2SO_4$ are diprotic acids, which give out two protons. Phosphoric acid ($H_3PO_4$) is a triprotic acid. In most cases, polyprotic acids do not fully dissociate and give out all the protons simultaneously. Dissociation constants for each dissociation varies. For example, in phosphoric first dissociation constant is $7.25\times10^{-33}$, which is a larger value. So the full dissociation takes place. Second dissociation constant is $6.31\times10^{-8}$ and the third one is $3.98\times10^{-13}$, which are less favorable dissociations than the first one.

The Table 5.1 gives values of $K_a$ for some common polyprotic acids. The large difference between the values of $K_a$ for the sequential loss of protons by a polyprotic acid is important because it means we can assume that these acids dissociate one step at a time as an assumption known as stepwise dissociation.

**Table 5.1: Acid-Dissociation Equilibrium Constants for Common Polyprotic Acids**

| Acid | $K_{a1}$ | $K_{a2}$ | $K_{a3}$ |
|---|---|---|---|
| Sulfuric acid ($H_2SO_4$) | $1.0 \times 10^3$ | $1.2 \times 10^{-2}$ | |
| Chromic acid ($H_2CrO_4$) | 9.6 | $3.2 \times 10^{-7}$ | |
| Oxalic acid ($H_2C_2O_4$) | $5.4 \times 10^{-2}$ | $5.4 \times 10^{-5}$ | |
| Sulfurous acid ($H_2SO_3$) | $1.7 \times 10^{-2}$ | $6.4 \times 10^{-8}$ | |
| Phosphoric acid ($H_3PO_4$) | $7.1 \times 10^{-3}$ | $6.3 \times 10^{-8}$ | $4.2 \times 10^{-13}$ |
| Glycine ($C_2H_6NO_2$) | $4.5 \times 10^{-3}$ | $2.5 \times 10^{-10}$ | |
| Citric acid ($C_6H_8O_7$) | $7.5 \times 10^{-4}$ | $1.7 \times 10^{-5}$ | $4.0 \times 10^{-7}$ |
| Carbonic acid ($H_2CO_3$) | $4.5 \times 10^{-7}$ | $4.7 \times 10^{-11}$ | |
| Hydrogen sulfide ($H_2S$) | $1.0 \times 10^{-7}$ | $1.3 \times 10^{-13}$ | |

Considering the consequence of assumption that polyprotic acids loses protons one step at a time by examining the chemistry of a saturated solution of $H_2S$ in water (Figure 5.1).

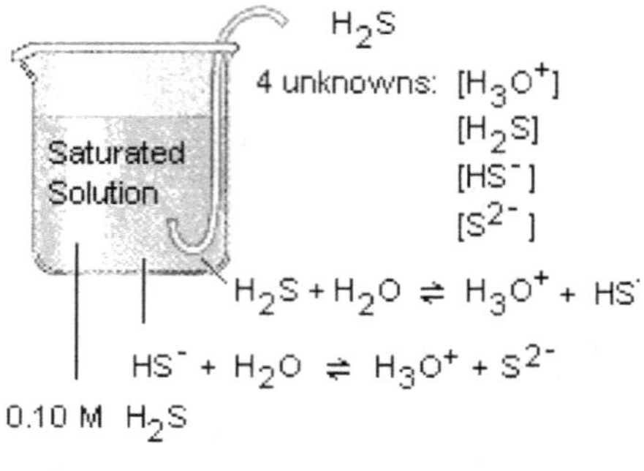


**Figure 5.1:** Chemistry of Saturated Solution of $H_2S$ in Water.

It is observed that Hydrogen sulfide is foul smelling gas that gives rotten eggs their unpleasant odor. It is an excellent source of the $S^{2-}$ ion, however, and is therefore commonly used in introductory chemistry laboratories. $H_2S$ is a weak acid that dissociates in steps.

Some of the $H_2S$ molecules lose a proton in the first step to form the $HS^-$, or hydrogen sulfide, ion.

First step: $H_2S(aq) + H_2O(l) \rightleftharpoons H_3O^+(aq) + HS^-(aq)$

A small fraction of the $HS^-$ ions formed in this reaction then go on to lose another $H^+$ ion in a second step.

Second step: $HS^-(AQ) + H_2O(L) \rightleftharpoons H_3O^+(AQ) + S^{2-}(AQ)$

Since there are two steps in this reaction, we can write two equilibrium constant expressions.

$$K_{al} = \frac{[H_3O^+][HS^-]}{[H_2S]} = 1.0 \times 10^{-7}$$

$$K_{a2} = \frac{[H_3O^+][S^{2-}]}{[HS^-]} = 1.3 \times 10^{-13}$$

As each of equations contains three terms, there are only four unknowns: $[H_3O^+]$, $[H_2S]$, $[HS^-]$, and $[S^{2-}]$ as the $[H_3O^+]$ and $[HS^-]$ terms appear in both equations. The $[H_3O^+]$ term shows total $H_3O^+$ ion concentration from both steps and therefore must have the same value in both equations. Also, $[HS^-]$ term that shows balance among $HS^-$ ions formed in first step and $HS^-$ ions consumed in second step, have similar value for both equations.

It is observed that four equations are required to solve for four unknowns, so with two equations: $K_{a1}$ and $K_{a2}$ expressions, we find either two more equations or pair of assumptions that can generate two equations. By assuming on fact that value of $K_{a1}$ for this acid is million times larger than value of $K_{a2}$

$$K_{a1} >> K_{a2}$$

It shows that only small fraction of $HS^-$ ions formed in first step will dissociate in second step. If this is true, most of the $H_3O^+$ ions in this solution come from the dissociation of $H_2S$, and most of the $HS^-$ ions formed in this reaction PSS remain in solution. With this, assume that the $H_3O^+$ and $HS^-$ ion concentrations are more or less equal.

First assumption: $[H_3O^+] \approx [HS^-]$

It is required to have one more equation, so one more assumption is necessary. Note that $H_2S$ is a weak acid ($K_{a1} = 1.0 \times 10^{-7}$, $K_{a2} = 1.3 \times 10^{-13}$) so assuming that most of $H_2S$ will dissolve in water that is present when solution reaches equilibrium. Further, it is assumed that equilibrium concentration of $H_2S$ is approximately equal to initial concentration.

Second assumption: $[H_2S] \approx C_{H2S}$

We now have four equations in four unknowns.

$$K_{al} = \frac{[H_3O^+][HS^-]}{[H_2S]} = 1.0 \times 10^{-7}$$

$$K_{a2} = \frac{[H_3O^+][S^{2-}]}{[HS^-]} = 1.3 \times 10^{-13}$$

$$[H_3O^+] \approx [HS^-]$$

$$[H_2S] \approx C_{H2S}$$

As there is always a unique solution to four equations in four unknowns, so calculating $H_3O^+$, $H_2S$, $HS^-$, and $S^{2-}$ concentrations at equilibrium in a saturated solution of $H_2S$ in water. Here it is required to have a saturated solution of $H_2S$ in water having initial concentration as 0.10 M.

As $K_{a1}$ is larger than $K_{a2}$ for acid, considering equilibrium expression for first step without worrying about second step, hence starting with expression for $K_{a1}$ for acid, gives

$$K_{al} = \frac{\left[H_3O^+\right]\left[HS^-\right]}{\left[H_2S\right]} = 1.0\times10^{-7}$$

Removing one of the assumptions

$[H_2S] \approx C_{H2S} \approx 0.10$ M

Substituting this approximation into the $K_{a1}$ expression gives the following equation.

$$\frac{\left[H_3O^+\right]\left[HS^-\right]}{\left[H_2S\right]} = 1.0\times10^{-7}$$

Removing other assumption

$[H_3O^+] \approx [HS^-] \approx \Delta C$

Substituting this approximation into the $K_{a1}$ expression gives the following result.

$$\frac{[\Delta C][\Delta C]}{[0.10]} \approx 1.0\times10^{-7}$$

We now solve this approximate equation for $\Delta C$.

$\Delta C \approx 1.0 \times 10^{-4}$

If two assumptions are valid, then $H_2S$, $H_3O^+$ and $HS^-$ concentrations will be:

$[H_2S] \approx 0.10$ M

$[H_3O^+] \approx [HS^-] \approx 1.0 \times 10^{-4}$ M

On extracting values of three unknowns from first equilibrium, turning second equilibrium expression as

$$K_{a2} = \frac{\left[H_3O^+\right]\left[S^{2-}\right]}{\left[HS^-\right]} = 1.3\times10^{-13}$$

Substituting known values of $H_3O^+$ and $HS^-$ ion concentrations in expression as

$$\frac{\left[1.0\times10^{-4}\right]\left[S^{2-}\right]}{\left[1.0\times10^{-4}\right]} = 1.3\times10^{-13}$$

As the equilibrium concentrations of $H_3O^+$ and $HS^-$ ions are more or less similar, then $S^{2-}$ion concentration at equilibrium is equal to value of $K_{a2}$ for this acid.

$[S^{2-}] \approx 1.3 \times 10^{-13}$ M

Now checking the assumptions, by comparing dissociation of $H_2S$ with initial concentration, so the $HS^-$ and $H_3O^+$ ion concentrations be obtained from calculation which is 0.1% of initial concentration of $H_2S$, hence following assumption is therefore valid.

$$[H_2S] \approx C_{H2S} \approx 0.10\ M$$

Now verify and analyse that difference between the $S^{2-}$ and $HS^-$ ion concentrations is large enough to allow to assume $H_3O^+$ ions at equilibrium that are formed in first step and $HS^-$ ions which are formed in next step in a solution. The $S^{2-}$ ion concentration obtained from this calculation is $10^9$ times smaller than the $HS^-$ ion concentration, hence assumption is valid.

$$[H_3O^+] \approx [HS^-]$$

Finally, the concentrations of various components of equilibrium will be:

$$[H_2S] \approx 0.10\ M$$

$$[H_3O^+] \approx [HS^-] \approx 1.0 \times 10^{-4}\ M$$

$$[S^{2-}] \approx 1.3 \times 10^{-13}\ M$$

## 5.4 Titration of Diprotic and Polyprotic Acids

### 5.4.1 Monoprotic Acids Titration

Monoprotic acids may be encountered during a chemical technique called a titration. A titration is a process in which you take a solution with a known concentration and use that to determine the solution with an unknown concentration Look at the relationship between monoprotic acids and this technique. A solution with a known concentration is added to the solution with the unknown concentration (Figure 5.2).

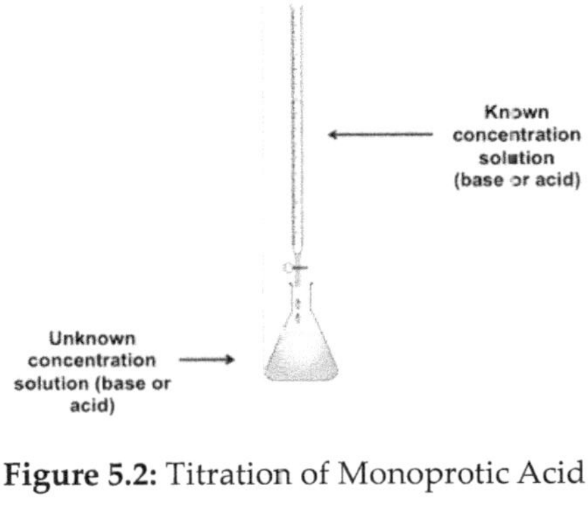


**Figure 5.2:** Titration of Monoprotic Acid.

Titrations, as seen in the diagram, provide a great method towards identifying monoprotic acids. The equivalence point in a titration is what clues you in to whether or not the given substance is monoprotic. An equivalence point is a stoichiometric value where the point in a titration corresponds to equal (or equivalent) amounts of acid and base being added. Don't be alarmed by the term stoichiometric. This simply means that the amount of the two solutions used in the titration are identical.

By graphing a titration curve, the equivalence point can be identified. A monoprotic acid will only have one equivalence point as monoprotic acid can only donate one acidic proton. In titration, a solution of sodium hydroxide (NaOH) is added to a solution of hydrogen chloride (HCl). At the equivalence point, the acidic proton, the hydrogen attached to chlorine, in hydrogen chloride reacts at an equivalent amount with the hydroxide ion from the sodium hydroxide.

There will be no net acidic protons or hydroxide ions left over. Because hydrogen chloride only has one acidic proton to donate, it is considered monoprotic. Recall that the number of equivalence points in a titration curve corresponds to the number of protons an acid can donate (Figure 5.3).

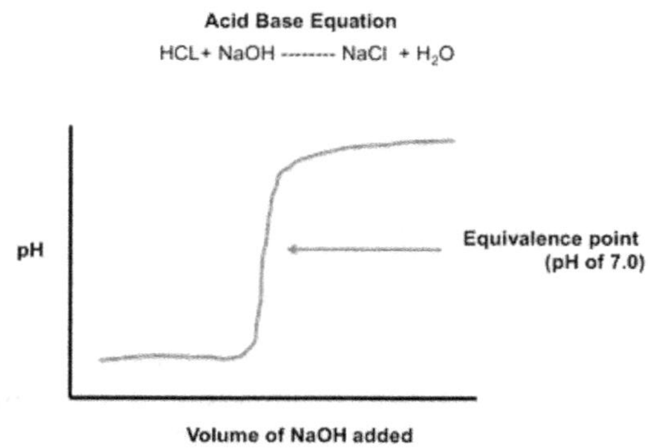


**Figure 5.3:** Graph of Titration Curve.

Given this concept, as well as the fact that hydrogen chloride can only donate one acidic proton, it makes perfect sense that this titration curve is monoprotic.

| Acid Name | Formula |
|---|---|
| Hydrochloric Achid | HCl |
| Nitiric Acid | $HNO_3$ |
| Acetic Acid | $CH_3COOH$ |
| Benzoic Acid | $C_6H_5COOH$ |
| Hydrolodic Acid | H |
| Hydroflurolc Acid | $H^+$ |

Examples

### 5.4.2 Titration of Polyprotic Acids

Titration of polyprotic acids or bases is slightly complex as compared to titration of monoprotic acid/base. In polyprotic acid, there appears more than one inflection point on titration curve. As in laboratory practice possible pH change during titration is limited to about 7-8 units at most, that means steep part of the titration curve - when split between two end points - must be short. As we have seen in the end point detection section it doesn't help to achieve good accuracy of titration.

Also, normally without using precise analysis, it is not obvious about stoichiometry of reaction. In such case, on titrating phosphoric acid with strong base, it is hard to observe third end point in titration curve so the curve is completely flat. Hence, it is required to choose an indicator that will allow titration of either only first proton or two and can calculate reaction stoichiometry accordingly. Titration curves of polyprotic acids, presented below, show these problems (Figure 5.4).

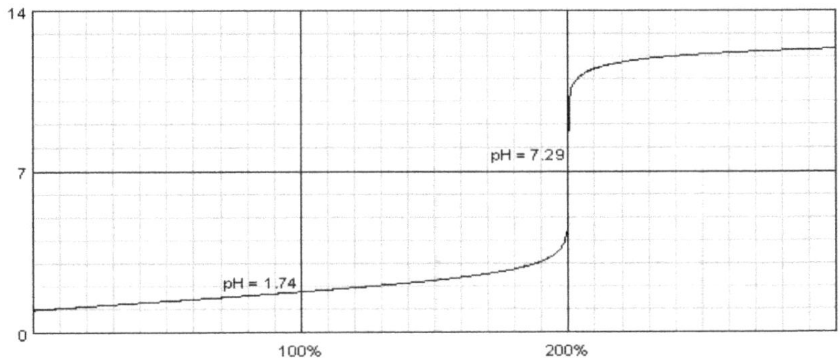


**Figure 5.4**: Titration Curves of Polyprotic Acids where 0.1M Solution of Sulfuric Acid is Titrated with 0.1M Solution of Strong Base.

It is observed from the curve that sulfuric acid where the second proton is much less acidic as compared to first one will be quiet strong enough so that both the protons should be titrated at the same time. In this cure, there results only single nice and high steep part of titration curve where all was told about indicator selection for titration of strong acids still holds (Figure 5.5).

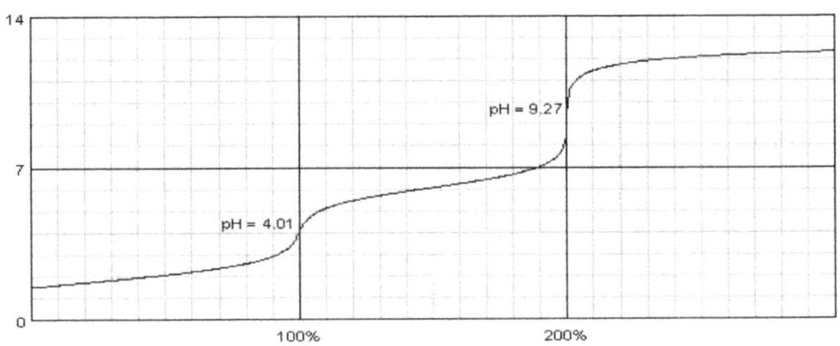


**Figure 5.5:** Titration Curves of Polyprotic Acids where 0.1M Solution of Maleic Acid Titrated with 0.1M Solution of Strong Base.

Carefully selecting both $pK_a$ values we can calculate better titration curve, with two almost identical end points. However, this is curve of a real substance - maleic acid. Curve at the first inflection point is not steep enough to be used for end point determination (Figure 5.6).

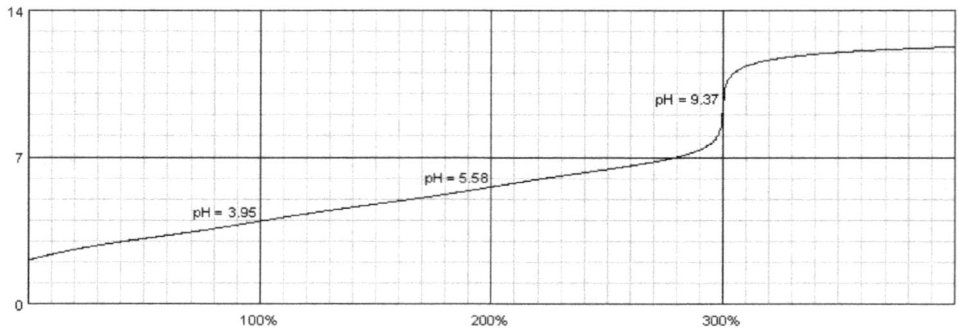


**Figure 5.6**: Titration Curves of Polyprotic Acids where 0.1M Solution of Citric Acid Titrated with 0.1M Solution of Strong Base

In the above curve, citric acid has three relatively similar dissociation constants, so instead of giving three or two separate end points, it has a long ramp where buffering effect of the first and second dissociation steps stops fast rise of pH (Figure 5.7).

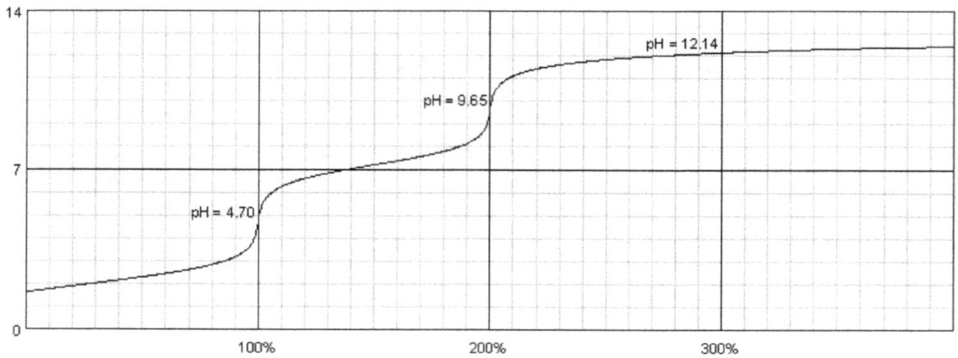


**Figure 5.7**: Titration Curves of Polyprotic Acids where 0.1M Solution of Phosphoric Acid Titrated with 0.1M Solution of Strong Base

In the case of phosphoric acid first two protons are similar in strength to protons in maleic acid, thus initially titration curve looks similar, with two separate end points. Third dissociation constant is so small, that even after adding large excess of 0.1M titrant over 20% of the acid is in the form of $HPO_4^{2-}$.

Note, that in almost all titration curves presented above, pH changes close to the end point are relatively slow compared to the titration curves of monoprotic acids (even weak, monoprotic acid). That means it is very difficult to precisely determine the end point of the titration.

While in many cases more then one end point makes titration difficult, it allows simultaneous determination of NaOH and $Na_2CO_3$ in one solution in so called Warder titration (Figure 5.8).

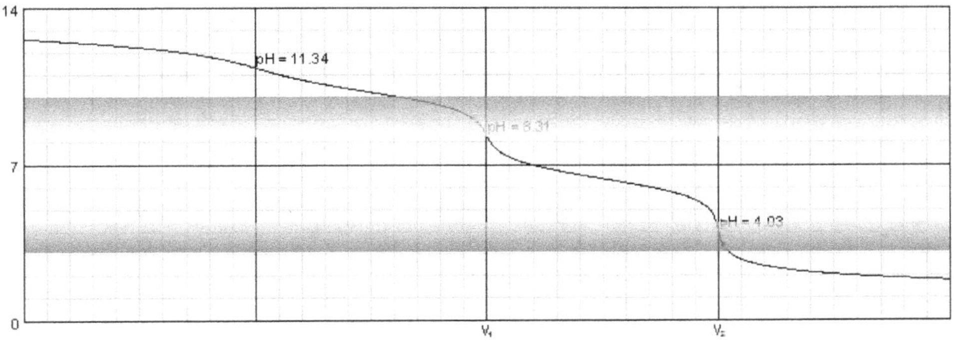


**Figure 5.8**: Warder Titration where Solution of 0.1M NaOH and 0.1M $Na_2CO_3$ Titrated with 0.1M Solution of Strong Acid.

Solution of sodium hydroxide and disodium carbonate contains three bases - $OH^-$, $CO_3^{2-}$ and $HCO_3^-$. First two are much stronger, so they are neutralized first. First equivalence point is at pH 8.31, almost exactly at pH where phenolphthalein starts to be visible. Second end point is at pH 4.03, right in the middle of methyl orange color change pH range.

During titration of the mixture we first add phenolphthalein to the solution and titrate it till pink color disappears, noting the first end point volume $V_1$. On adding methyl orange and titrate solution to intermediate color of indicator, the second end point volume be considered as $V_2$.

$V_1$ is the volume of titrant that was necessary for completion of two reactions:

$$NaOH + HCl \rightarrow NaCl + H_2O$$

$$Na_2CO_3 + HCl \rightarrow NaCl + NaHCO_3$$

$V_2$-$V_1$ is the hydrogen carbonate protonation:

$$NaHCO_3 + HCl \rightarrow NaCl + H_2O + CO_2$$

Obviously $V_2$-$V_1$ can be directly used to calculate amount of the carbonate present in the original solution. $V_1$ reflects a sum of NaOH and $Na_2CO_3$. As we know $V_2$-$V_1$ is the volume needed for $Na_2CO_3$ protonation, we can easily calculate that $2V_1$-$V_2$ is the volume of the titrant necessary to neutralize just NaOH.

### 5.4.3 Titration of Diprotic Acids/Bases

Diprotic acid is an acid which gives two $H^+$ ions per acid molecule. The examples of diprotic acids are sulfuric acid $H_2SO_4$, and carbonic acid $H_2CO_3$. Diprotic acid dissociates in water in two stages:

$$H_2X(aq) \rightleftharpoons H^+_{(aq)} + HX^-_{(aq)}$$

$$HX^-_{(aq)} \rightleftharpoons H^+_{(aq)} + X^{2-}_{(aq)}$$

Due to successive dissociations, titration curves of diprotic acids have two equivalence points as shown in Figure 5.9.

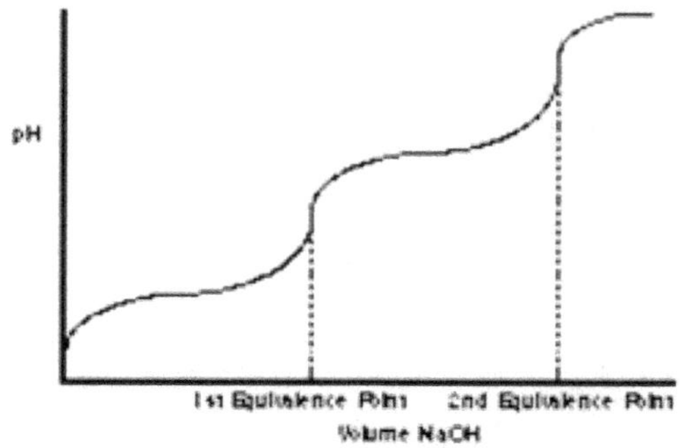


**Figure 5.9:** Successive Dissociations.

The equations for the acid-base reactions occurring between a diprotic acid $H_2X$, and sodium hydroxide base NaOH, are:

$$H_2X + NaOH \rightarrow NaHX + H_2O$$

from the beginning to the first equivalence point:

from the first to the second equivalence point:

$$NaHX + NaOH \rightarrow Na_2X + H_2O$$

from the beginning of the reaction through the second equivalence point (net reaction):

$$H_2X + 2\ NaOH \rightarrow Na_2X + 2\ H_2O$$

At first equivalence point, all $H^+$ ions from first dissociation reacts with NaOH base. At the second equivalence point, all H+ ions from both reactions reacts. Hence volume of NaOH added at the second equivalence point will be twice that of first equivalence point from above equations.

Diprotic acid when titrated with NaOH solution of known concentration result in molecular weight in g/mole. Weighing sample of acid will allow to have known with the mass in grams. Moles can be obtained from volume of NaOH titrant that is required to reach the first equivalence point. The volume and concentration of

NaOH titrant calculates moles of NaOH. Moles of unknown acid equal moles of NaOH at first equivalence point. After knowing the grams and moles of diprotic acid, molecular weight be calculated in g/mole. Molecular weight determination is a common way of identifying an unknown substance, so in this case, first or second equivalence point can be utilized to find molecular weight.

Here, first will be easier as moles of NaOH are equal to moles of $H_2X$. If second equivalence point is clearly defined on titration curve, then divide its NaOH volume by 2 to confirm the first equivalence point or use the ratio:

1 mole $H_2X$ / 2 mol NaOH

Using half-titration method, it is possible to find the acid dissociation constants $K_{a1}$ and $K_{a2}$, for the two dissociations of diprotic acid. The Ka expressions for first and second dissociations from above equations results as:

$$K_{a1} = \frac{[H^+][HX^-]}{[H_2X]} \qquad K_{a2} = \frac{[H^+][X^{2-}]}{[HX^-]}$$

In this, first half-titration point occurs when one-half of $H^+$ ions in first dissociation gets titrated with NaOH resulting as:

$[H_2X] = [HX^-]$

Similarly second half titration point occurs when one-half of $H^+$ ions in second dissociation have been titrated with NaOH, resulting as:

$[HX^-] = [X^{2-}]$

On substituting $[H_2X]$ for $[HX^-]$ in $K_{a1}$ expression while $[HX^-]$ for $[X^{2-}]$ in $K_{a2}$ expressions, then following results obtained as:

$K_{a1} = [H^+]$ $K_{a2} = [H^+]$

Taking base-ten log both sides of equation,

$\log K_{a1} = \log[H^+]$

$\log K_{a2} = \log[H^+]$

following expressions are obtained:

$pK_{a1} = pH$

$pK_{a2} = pH$

So pH value at first half-titration volume Point 1 is shown in figure which is equal to $pK_{a1}$ value. In this, first half-titration point volume be found by dividing first equivalence point volume by two.

Likewise, pH value at second titration point will be equal to the $pK_{a2}$ value and second half titration volume Point 2 will be the midpoint between first and second equivalence point volumes (EP and EP) (Figure 5.10).

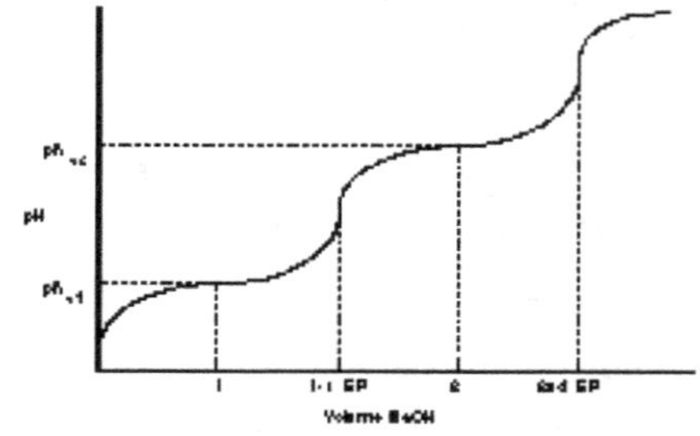

**Figure 5.10**: Titration Curve.

### 5.4.4 Diprotic Bases

The techniques used in titrating for diprotic acids can be applied to diprotic bases where it is required to find the values of $K_b$ for base.

Now we will find the $H_2CO_3$, $HCO_3^-$, $CO_3^{2-}$, and $OH^-$ concentrations at equilibrium in a solution that is initially 0.10 M in $Na_2CO_3$. ($H_2CO_3$: $K_{a1} = 4.5 \times 10^{-7}$; $K_{a2} = 4.7 \times 10^{-11}$)

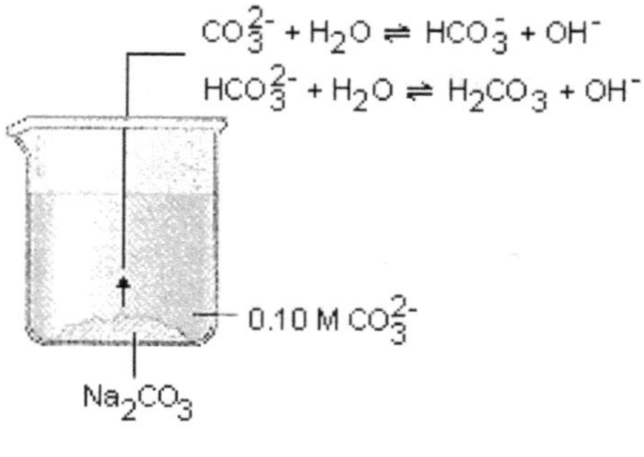


**Figure 5.11**: Concentrations at Equilibrium in Solution.

As it is a salt, sodium carbonate dissociates into its ions when it dissolves in water.

$$Na_2CO_3(aq) \xrightarrow{H_2O} 2\,Na^+(aq) + CO_3^{2-}(aq)$$

The carbonate ion then acts as a base toward water, picking up a pair of protons (one at a time) to form the bicarbonate ion, $HCO_3^-$ ion, and then eventually carbonic acid, $H_2CO_3$.

$$CO_3^{2-}(aq) + H_2O(l) \rightleftharpoons HCO_3^-(aq) + OH^-(aq) \qquad K_{b1} = ?$$

$$HCO_3^-(aq) + H_2O(l) \rightleftharpoons H_2CO_3(aq) + OH^-(aq) \qquad K_{b2} = ?$$

The first step in solving this problem involves determining the values of $K_{b1}$ and $K_{b2}$ for the carbonate ion. Now compare $K_b$ expressions for carbonate ion with $K_a$ expressions for carbonic acid.

$$K_{b1} = \frac{[HCO_3^-][OH^-]}{[CO_3^{2-}]} \qquad K_{a2} = \frac{[H_3O^+]}{[HCO_3^-]}$$

$$K_{b2} = \frac{[H_2CO_3][OH^-]}{[HCO_3^-]} \qquad K_{al} = \frac{[H_3O^+][HCO_3^-]}{[H_2CO_3]}$$

The expressions for $K_{b1}$ and $K_{a2}$ have something in common — they both depend on the concentrations of the $HCO_3^-$ and $CO_3^{2-}$ ions. The expressions for $K_{b2}$ and $K_{a1}$ also have something in common — they both depend on the $HCO_3^-$ and $H_2CO_3$ concentrations. We can therefore calculate$K_{b1}$ from $K_{a2}$ and $K_{b2}$ from $K_{a1}$.

Further multiply top and bottom of $K_{a1}$ expression by $OH^-$ ion concentration to have $[OH^-]$ term.

$$K_{al} = \frac{[H_3O^+][HCO_3^-]}{[H_2CO_3]} \times \frac{[OH^-]}{[OH^-]}$$

Now group the terms in equation

$$K_{al} = \frac{[H_3O^+][HCO_3^-]}{[H_2CO_3]} \times [H_3O^+][OH^-]$$

Here, first term in equation is inverse of $K_{b2}$ expression, and second term is $K_w$expression

$$K_{al} = \frac{1}{K_{b2}} \times K_w$$

Rearranging this equation gives the following result.

$K_{a1}K_{b2} = K_w$

Further, multiply top and bottom of $K_{a2}$ expression by $OH^-$ion concentration

$$K_{a2} = \frac{[H_3O^+][CO_3^{2-}]}{[HCO_3^-]} \times \frac{[OH^-]}{[OH^-]}$$

Collect terms in the equation

$$K_{a2} = \frac{\left[H_3O^+\right]\left[CO_3^{2-}\right]}{\left[HCO_3^-\right]} \times \left[OH^-\right]\left[H_3O^+\right]$$

The first term in this equation is the inverse of $K_{b1}$, and the second term is $K_w$.

$$K_{a2} = \frac{1}{K_{b1}} \times K_w$$

This equation can therefore be rearranged as follows.

$$K_{a2}K_{b1} = K_w$$

Now find the values of $K_{b1}$ and $K_{b2}$ for carbonate ion:

$$K_{b1} = \frac{K_w}{K_{a2}} = \frac{1.0\times10^{-14}}{4.7\times10^{-11}} = 2.1\times10^{-4}$$

$$K_{b2} = \frac{K_w}{K_{a1}} = \frac{1.0\times10^{-14}}{4.5\times10^{-7}} = 2.2\times10^{-8}$$

For calculating, consider $K_{b1}$ expression as the $CO_3^{2-}$ ion is strongest base in solution and hence the best source of $OH^-$ ion, so

$$K_{b1} = \frac{\left[HCO_3^-\right]\left[OH^-\right]}{\left[CO_3^{2-}\right]} = 2.1\times10^{-4}$$

Here, the difference between $K_{b1}$ and $K_{b2}$ for carbonate ion is large enough to suggest that most of the $OH^-$ ions come from this step and many $HCO_3^-$ ions formed in reaction remain in solution.

$$[OH^-] \approx [HCO_3^-] \approx \Delta C$$

Further, value of $K_{b1}$ is small enough to think that $\Delta C$ is small as compared with initial concentration of carbonate ion. If it is correct then concentration of $CO_3^{2-}$ ion at equilibrium will be equal to initial concentration of $Na_2CO_3$

$$[CO_3^{2-}] \approx C_{Na2CO3}$$

On substituting above information in $K_{b1}$ expression:

$$\frac{[\Delta C][\Delta C]}{[0.10]} \approx 2.1\times10^{-4}$$

This approximate equation can now be solved for ΔC.

$$\Delta C \approx 0.0046 \text{ M}$$

Using the value of ΔC and finding equilibrium concentrations of $OH^-$, $HCO_3^-$, and $CO_3^{2-}$ ions

$$[CO_3^{2-}] = 0.10 - \Delta C \approx 0.095 \text{ M}$$

$$[OH^-] \approx [HCO_3^-] \approx \Delta C \approx 0.0046 \text{ M}$$

Considering the $K_{b2}$ expression

$$K_{b2} = \frac{[H_2CO_3][OH^-]}{[HCO_3^-]}$$

Putting information about $OH^-$ and $HCO_3^-$ ion concentrations in equation

$$\frac{[H_2CO_3][0.0046]}{[0.0046]} \approx 2.2 \times 10^{-8}$$

The above equation shows that $H_2CO_3$ concentration at equilibrium is equal to $K_{b2}$ for carbonate ion.

$$[H_2CO_3] \approx 2.2 \times 10^{-8} \text{ M}$$

Finally, the calculations allows to test the assumptions made generating these results as:

$$[CO_3^{2-}] \approx 0.095 \text{ M}$$

$$[OH^-] \approx [HCO_3^-] \approx 4.6 \times 10^{-3} \text{ M}$$

$$[H_2CO_3] \approx 2.2 \times 10^{-8} \text{ M}$$

As all the assumptions are correct, so extent of the reaction between $CO_3^{2-}$ ion and water will give $HCO_3^-$ ion which is less than 5% of initial concentration of $Na_2CO_3$. Furthermore, many $OH^-$ ion appears from first step and many $HCO_3^-$ ion formed will remain in solution.

## 5.5 Conjugate Base of Monoprotic Acid

A conjugate base contains one less H atom and one more - charge than the acid that formed it. Let us take the example of bicarbonate ions reacting with water to create carbonic acid and hydronium ions.

$$HCO_3^- + H_2O \rightarrow H_2CO_3 + OH^-$$

base + acid → Conj A + Conj B

It is observed that $HCO_3^-$ becomes $H_2CO_3$ having one more H atom and one more + charge (-1 + 1 = 0). Hence $H_2CO_3$ is the conjugate acid of $HCO_3^-$. The $H_2O$ becomes $OH^-$. It has one less H atom and one more – charge. So $OH^-$ is the conjugate base of $H_2O$.

# 6 Solubility

## 6.1 Introduction

Solution is a mixture of two or more substances. Solute is a substance being dissolved, and the solvent is substance where solute is disscived in. The solvent is often water. The process of creating a solution is called solvation. Solutions are homogeneous mixtures of solvents (the larger volume of the mixture) and solutes (the smaller volume of the mixture). The solute and solvent can be either a solid, liquid or a gas. A solution forms when the attractive forces between the solute and the solvent are similar (Figure 6.1).

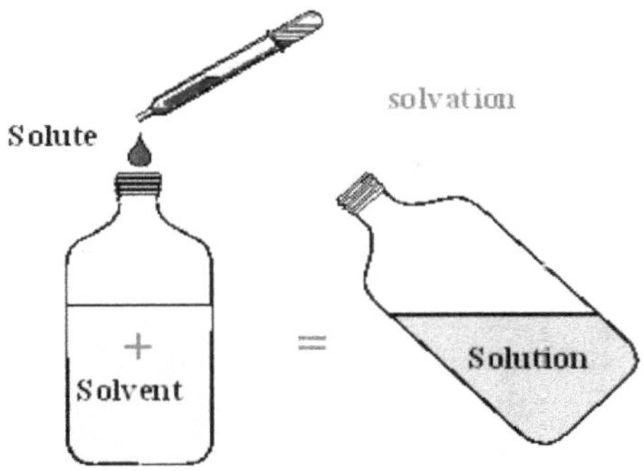


**Figure 6.1:** Solution.

Solubility is the quantity of solute that dissolves in a given amount of solvent. The solubility of a solute depends on the nature of the solute and solvent, the amount of solute, the temperature and pressure (for a gas) of the solvent. Solubility is expressed as the quantity of solute per 100 g of solvent at a specific temperature.

It is observed that everything cannot be dissolve in water such as oil, which cannot dissolve in water. Whether or not a solute will dissolve in a particular solvent depends on the attraction between molecules. One rule to remember like dissolves like, that shows that ionic or polar solutes will dissolve in similar polar solvents. Non-polar solutes dissolve in non-polar solvents. However, non-polar solutes, such as oil, will not dissolve in polar solvents, such as water, and vice-versa. As with many rules, there are exceptions in certain cases.

When a crystal of salt is placed in water, which is then stirred, the crystal dissolves and a clear solution is formed. When more salt is added to salt-water solution, it too dissolves, making the solution more concentrated than the previous one. Even more salt can be dissolved in the water to make it a much more concentrated salt solution. Thus, one of properties of solutions is that they have a variable composition.

When salt dissolved in water to make salt water, the solution formed is clear and colorless.

As the water molecules collide with the ionic compound (NaCl), the charged ends of the water molecule become attracted to the positive sodium ions and negative chloride ions. The water molecules surround the ions and the ions move into solution. This process of attraction between the water molecules (the solvent) and the ionic compound (NaCl, the solute) is called solvation. Solvation continues until the entire crystal has dissolved and all ions are distributed throughout the solvent (Figure 6.2).

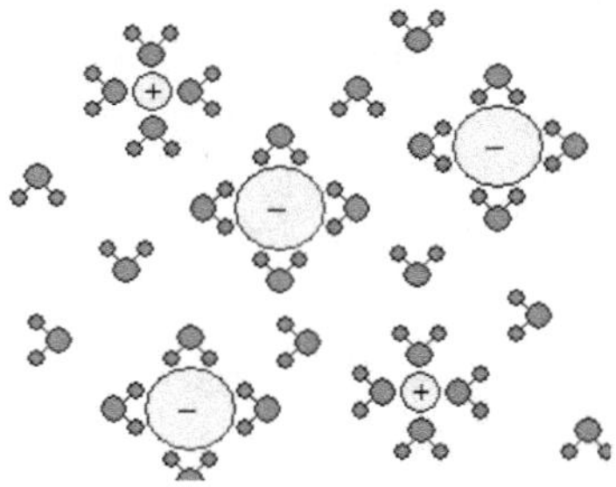

**Figure 6.2**: Solvation Process of NaCl.

When sugar is dissolved in water, again a clear, colorless solution is formed. When copper sulfate is dissolved in water, it also forms a clear solution. However, the solutions formed with the salt and the sugar are colorless, whereas that formed with cooper sulfate is blue. Solutions are always clear. They may or may not have a color. Clear merely means that the solution is transparent to light.

When a solution is allowed to stand undisturbed for a long period of time, no crystals of solute settle out, provided the solvent is not allowed to evaporate. This is another property of solutions – the solute not settle out.

The salt in salt – water solution can be recovered by allowing the water to evaporate; the same is true of the sugar in a sugar solution. In general, solutions can be separated by physical means.

### 6.1.1 Properties of Solutions

- Solutions consist of s soluble materials (the solute) dissolved in a liquid (the solvent)
- Solutions have a variable composition

- Solutions are clear
- Solution are homogeneous
- Solutions can be separated by physical means

### 6.1.2 Precipitation

After a solution is created, solid can be formed from it. This occurs in liquid solution where the formation of solid is known as precipitate while remaining liquid in solution is known as supernate. Sometimes, a rotating machine called a centrifuge can be used to help collect all of the small particles that are still floating around in the liquid, called the suspension (Figure 6.3).

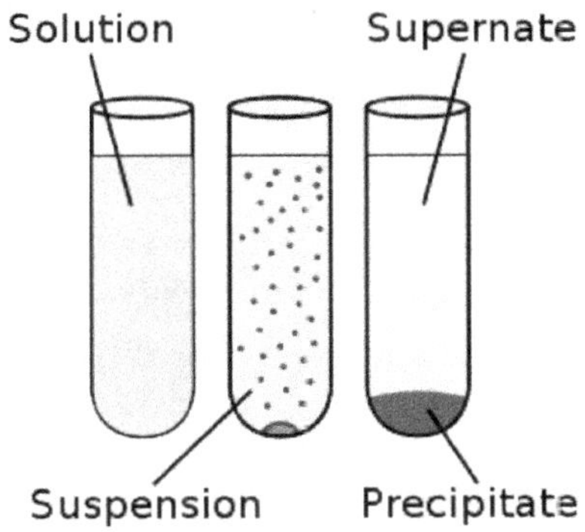


**Figure 6.3**: Precipitation of Solution.

### 6.1.3 Application

A technique which scientists can use to concentrate DNA or RNA is ethanol precipitation. In an aqueous solution of DNA or RNA, salt and ethanol are added to solution where nucleic acid precipitates out of it. The solution is further centrifuged and nucleic acid forms a pellet separate from the solution which is washed in ethanol and then centrifuged again and further dries up and is re-suspended in clean aqueous buffer.

As water is polar due to its positive hydrogen and negative oxygen, then nucleic acids like DNA and RNA also results as polar due to negatively charged phosphate group ($PO_3^-$) along the sugar-phosophate backbone. Due to this, DNA and RNA can easily dissolve in water.

Adding of salt will neutralize charges on sugar-phosphate backbone which breaks into positive and negative ions where positive ion ($Na^+$) neutralizes negative phosphate group ($PO_3^-$), which makes the molecule less polar and less soluble in water. Addition of ethanol will helps $Na^+$ and $PO_3^-$ ions to come closer due to ions travel easier in ethanol as compared to water. The supernate is removed and new ethanol is added in the second washing. This removes any residual salt that remained on the pelleted DNA (Figure 6.4).

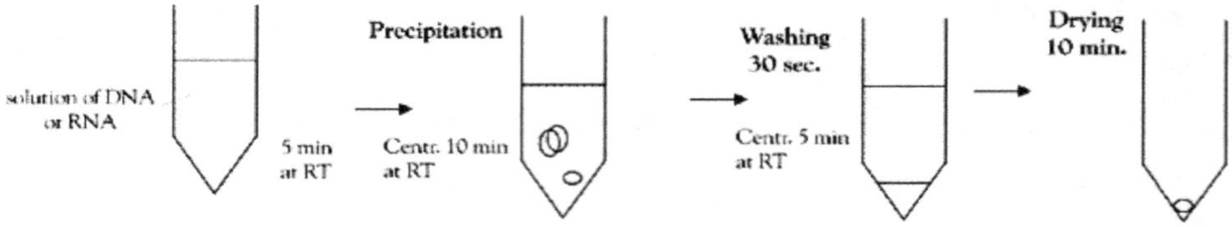


Figure 6.4

### 6.1.4 Rules

- All nitrates and acetates are soluble.
- All salts of group I ($Li^{1+}$ ◎ $Cs^{1+}$) and $NH_4^{1+}$ are soluble.
- All halides ($F^{1-}$ ◎ $I^{1-}$) are soluble, except those of $Ag^{1+}$, $Hg_2^{2+}$, and $Pb^{2+}$.
- All sulfates are soluble, except those of $Ca^{2+}$ (slightly), $Sr^{2+}$, $Ba^{2+}$, $Pb^{2+}$, and $Hg_2^{2+}$.
- All hydroxides, phosphates, carbonates and sulfides are insoluble. Hydroxides of $Ca^{2}$, $Sr^{2+}$, and $Ba^{2+}$ are *slightly* soluble.
- Compounds not covered above are probably not soluble.

| $NO_3^-$ | All soluble |
|---|---|
| $ClO_4^-$ | All soluble |
| $C_2H_3O_2^-$ | All soluble |
| $Cl^-$, $Br^-$, $I^-$ | All soluble except $Ag^{+1}$, $Hg_2^{+2}$, $Pb^{2+}$ |
| $SO_4^{-2}$ | All soluble except $Ba^{2+}$, $Sr^{2+}$, $Pb^{2+}$, $Hg_2^{+2}$ |
| $CO_3^{2-}$, $PO_4^{3-}$ | ONLY Group 1 (Alkali Metals), and $NH_4^+$ are soluble |
| $OH^-$ | ONLY Group 1 (Alkali Metals), $Sr^{2+}$, $Ba^{2+}$ and $NH_4^+$ are soluble |
| $S^{2-}$ | ONLY Group 1 (Alkali Metals), $Sr^{2+}$, $Ba^{2+}$ and $NH_4^+$ are soluble |

## 6.2 Solubility Equilibrium

Solubility equilibrium is base on the assumption that solids dissolve in water to give the basic particles from which they are formed. Molecular solids dissolve to give individual aqueous molecules.

$$C_{12}H_{22}O_{11(s)} \xrightarrow{H_2O} C_{12}H_{22}O_{11(aq)}$$

Ionic solids dissociate to give their respective positive and negative ions:

$$NaCl_{(s)} \xrightarrow{H_2O} Na^+_{(aq)} + Cl^-_{(aq)}$$

The ions in formed from the dissociation of ionic solids can carry an electrical current. Salt solutions, therefore, are good conductors of electricity. Molecular solids, however, do not dissociate in water to give ions, so no electrical current can be carried.

**Solubility**

Solubility is the ratio of maximum amount of solute to the volume of solvent in which solute can dissolve. It is expressed as grams of solute per 100 g of water and moles of solute per Liter of solution.

**Soluble**

These can be dissolve and will not form a solid precipitate. The examples are Alkali metal ions and ammonium ion: $Li^+$, $Na^+$, $K^+$, $NH_4^+$

acetate ion: $C_2H_3O_2^{1-}$

nitrate ion: $NO_3^{1-}$

halide ions (X): $Cl^-$, $Br^-$, $I^-$ (Exceptions: AgX, HgX, and $PbX_2$ are insoluble)

sulfate ion: $SO_4^{2-}$ (Exceptions: $SrSO_4$, $BaSO_4$, and $PbSO_4$ are insoluble;

$AgSO_4$, $CaSO_4$, and $Hg_2SO_4$ are slightly soluble)

**Insoluble**

These do not dissolve and forms a solid precipitate.

carbonate ion: $CO_3^{2-}$

chromate ion: $CrO_4^{2-}$

phosphate ion: $PO_4^{3-}$

sulfide ion: $S^{2-}$ (Exceptions: CaS, SrS, and BaS are soluble)

hydroxide ion: $OH^-$ (Exceptions: $Sr(OH)_2$ and $Ba(OH)_2$ are soluble;

$Ca(OH)_2$ is slightly soluble)

A salt is considered soluble if it dissolves in water to give a solution with a concentration of at least 0.1 M at room temperature.

A salt is considered insoluble if the concentration of an aqueous solution is less than 0.0001 M at room temperature.

Salts with solubilities between 0.0001 M and 0.1 M are considered to be slightly soluble.

Salts that have extremely low solubilities dissociate in water according to the principles of equilibrium. For example, the reaction for the dissociation of the salt AgCl is:

$$AgCl_{(s)} \xrightarrow[\text{dissociate}]{\text{dissolve}} Ag^+_{(aq)} + Cl^-_{(aq)}$$

The reverse reaction for the dissolving of the salt would be the precipitation of the ions to form a solid:

$$Ag^+_{(aq)} + Cl^-_{(aq)} \xrightarrow[dissociate]{preciitate} AgCl_{(s)}$$

The system has reached equilibrium when the rate at which AgCl dissolves is equal to the rate at which AgCl precipitates.

Considering the rules of solubility:

- Salts are generally more soluble in HOT water (Gases are more soluble in COLD water)
- Alkali Metal salts are very soluble in water, NaCl, KOH, $Li_3PO_4$, $Na_2SO_4$ etc...
- Ammonium salts are very soluble in water. $NH_4Br$, $(NH_4)_2CO_3$ etc.
- Salts containing the nitrate ion, $NO_3^-$, are very soluble in water.
- Most salts of $Cl^-$, $Br^-$ and $I^-$ are very soluble in water - exceptions are salts containing $Ag^+$ and $Pb^{2+}$.
    - soluble salts: $FeCl_2$, $AlBr_3$, $MgI_2$ etc.
    - insoluble salts: AgCl, $PbBr_2$ etc.

**Saturated Solution**

Solution which has maximum concentration of ions which exists in equilibrium with solid salt at a given temperature is known as saturated solution. The equilibrium reaction for dissociation of AgCl is:

$$AgCl_{(s)} \xrightleftharpoons{H_2O} Ag^+_{(aq)} + Cl^-_{(aq)}$$

In saturated solution, there will be no change in amount of solid precipitate at bottom of the beaker. In this, the concentration of the solution is constant where the rate at which the salt is dissolving into solution will be equal to the rate of precipitation.

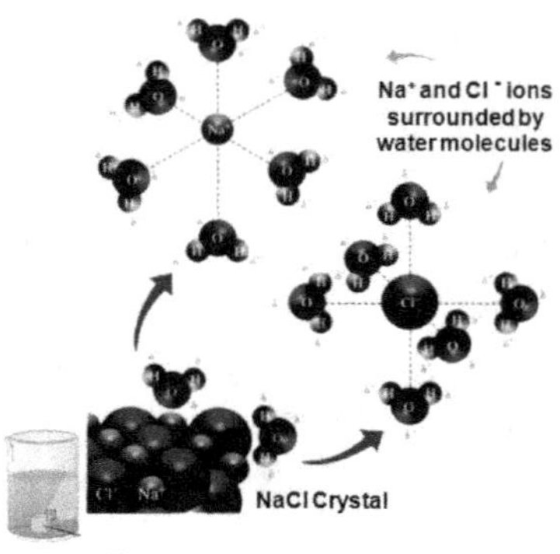


**Figure 6.5:** Salt Solution.

## Dissolving of Salt

A salt is an ionic compound - usually a metal cation bonded to a non-metal anion. The dissolving of a salt is an example of equilibrium.

**Figure 6.6**: Salt Solution in Water

The cations and anions are attracted to each other in the salt. They are also attracted to the water molecules. The water molecules will start to pull out some of the ions from the salt crystal.

At first, the only process occurring is the dissolving of the salt - the dissociation of the salt into its ions.

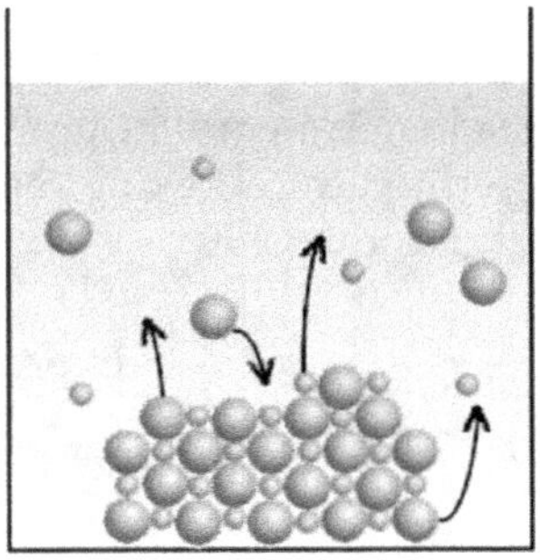

**Figure 6.7**: Dissociation of the Salt into its Ions.

However, soon the ions floating in the water begin to collide with the salt crystal and are "pulled back in" to the salt. (precipitation)

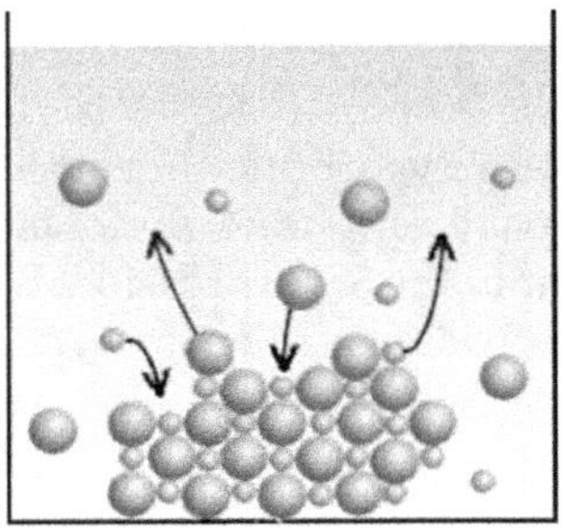

**Figure 6.8**: Rate of Precipitation of Salt Solution.

Eventually the rate of dissociation is equal to the rate of precipitation. The solution is now "saturated". It has reached equilibrium.

**Dissolving Silver Sulphate, $Ag_2SO_4$, in Water**

- When silver sulfate dissolves it dissociates into ions. When the solution is saturated, the following equilibrium exists: $Ag_2SO_4$ (s) D 2 $Ag^+$ (aq) + $SO_4^{2-}$ (aq)
- Since this is an equilibrium, we can write an equilibrium expression for the reaction: Ksp = $[Ag^+]^2[SO_4^{2-}]$
- Notice that the $Ag_2SO_4$ is left out of the expression. Since K is always calculated by just multiplying concentrations, it is called a solubility product constant - Ksp.

### 6.2.2 Solubility Product Expressions

Solubility product is an equilibrium constant that shows maximum amount of solid that can be dissolved in aqueous solution. Salts like Agl, BaS04, PbS04, Pbl2, etc, are ordinarily considered insoluble but they do possess some solubility. These are sparingly soluble electrolytes. A saturated solution of sparingly soluble electrolyte contains a very small amount of the dissolved electrolyte. It is assumed that whole of the dissolved electrolyte is present in the form of ions, i.e., it is completely dissociated.

Formations of precipitates are chemical equilibria phenomena, so heterogeneous equilibrium can normally be written in following manner, which further calls as equilibrium constants solubility products, $K_{sp}$:

| Equilibrium | Expression for $K_{sp}$ and $Q_{sp}$ |
|---|---|
| $AgCl(s) = Ag^+(aq) + Cl^-(aq)$ | $[Ag^+]\,[Cl^-]$ |
| $CaCO_3(s) = Ca^{2+}(aq) + CO_3^{2-}(aq)$ | $[Ca^{2+}]\,[CO_3^-]$ |
| $Li_2CO_3(s) = 2\,Li^+(aq) + CO_3^{2-}(aq)$ | $[Li^+]^2\,[CO_3^-]$ |

Solubility product, $K_{sp}$, is a special type of equilibrium constant which is given to solution having less soluble salts. Here, (aq) shows that ions are surrounded by water molecules which are in solution.

Also, the expression for solubility product are special equilibrium constant as solid present has fixed tendency of getting dissolved. Further, the role in $K_{sp}$ is constant as they do not appear in the $K_{sp}$ expression. If solution is not saturated, no precipitate will form and the product will be known as ion product, $Q_{sp}$.

$Q_{sp}$, $K_{sp}$ and Saturation

In certain substances, formation of solid or crystallization will not take place automatically whenever a solution is saturated. These substances results in tendency of forming oversaturated solutions. In case of syrup and honey, these are oversaturated sugar solutions having different substances like citric acids. In oversatureated solutions, $Q_{sp}$ is more than $K_{sp}$. When a seed crystal is provided or formed, a precipitate will form immediately due to equilibrium of requiring $Q_{sp}$ to approach $K_{sp}$.

Sodium acetate trihydrate, $NaCH_3COO{\cdot}3H_2O$, when heated to 370 K will become a liquid. The sodium acetate is said to be dissolved in its own water of crystallization. The substance stays as a liquid when cooled to room temperature or even below 273 K. As soon as a seed crystal is present, crystallization occur rapidly. In such a process, heat is released, and the liquid feels warm. Thus, the relationship among $Q_{sp}$, $K_{sp}$ and saturation is given below:

| | |
|---|---|
| $Q_{sp} < K_{sp}$ | Unsaturated solution |
| $Q_{sp} = K_{sp}$ | Saturate solution |
| $Q_{sp} > K_{sp}$ | Oversaturate solution |

**Molar Solubility and Solubility Products**

Solubility products, $K_{sp}$, of salts are indirect indication of their solubilities expressed in mol/L (called **molar solubility**). However, the solubility products are more useful than molar solubility. The molar solubilities are affected when there are common ions present in the solution. We need to employ the solubility products to estimate the molar solubilities in these cases.

When a salt is dissolved in pure water, solubility products and molar solubilities are related. This is illustrated using calcium carbonate. If $x$ is the concentration of $Ca^{2+}$ (= $[CO_3^{2-}]$) in the saturated solution, then

$$K_{sp} = x^2$$

In this case, $x$ is also called the molar solubility of $CaCO_3$. The following examples illustrate the relationship between solubility products, $K_{sp}$, and molar solubilities.

**Different Types of Expressions**

**Electrolyte of Type AB**

Its ionisation is represented as $AB \rightleftharpoons A^+ + B^-$

$$K_s = [A^+][B^-]$$

$$AgCl \rightleftharpoons Ag^+ + Cl^-;$$

$$K_s = [Ag^+][Cl^-]$$

$$BaSO_4 \rightleftharpoons Ba^{2+} + SO^{-2}_4$$

$$K_s = [Ba^{2+}][SO^{-2}_4]$$

### Electrolyte of Type $AB_2$

Its ionisation is represented as $AB_2 \rightleftharpoons A^{2+} + 2B^-$

$K_s = [A^{2+}][B^-]^2$

$PbCl_2 \rightleftharpoons Pb^{2+} + 2Cl^-$

$K_s = [Pb^{2+}][Cl^-]^2$

$CaF_2 \rightleftharpoons Ca^{2+} + 2F^-$

$K_s = [Ca^{2+}][F^-]^2$

### Electrolyte of Type $A_2B$

Its ionisation is represented as $A_2B \leftrightarrow 2A^{2+} + B^{2-}$

$K_s = [A^+]^2[B^{2-}]$

$Ag_2CrO_4 \rightleftharpoons 2Ag^+ + CrO^{-2}{}_4$

$K_s = [Ag^+]^2[CrO^{-2}{}_4]$

$H_2S \rightleftharpoons 2H^+ + S2^-$

$K_s = [H^+]^2[S^{2-}]$

### Electrolyte of Type $A_2B_3$

Its ionisation is represented as $A_2B_3 \rightleftharpoons 2A^{3+} + 3B^{2-}$

$K_s = [A^{3+}]^2[B^{2-}]^3$

$As_2S_3 \rightleftharpoons 2As^{3+} + 3S^{2-}$

$K_s = [As^{3+}]^2[S^{2-}]^3$

$Sb_2S_3 \rightleftharpoons 2Sb^{3+} + 3S2^-$

$K_s = [Sb^{3+}]^2[S^{2-}]^3$

### Electrolyte of Type $AB_3$

Its ionisation is represented as $AB_3 \rightleftharpoons A^{3+} + 3B^-$

$K_s = [A^{3+}][B^{2-}]^3$

$Fe(OH)_3 \rightleftharpoons Fe^{3+} + 3OH^-$

$K_s = [Fe^{3+}][OH^-]^3$

$AH_3 \rightleftharpoons Al^{3+} + 3l^-$

$K_s = [Al^{3+}][I^-]^3$

**Solubility Product of Weak Electrolyte**

Let degree of ionization of weak electrolyte $A_mB_n$ be ‹$\alpha$›.

$A_mB_n \rightleftharpoons mA^{n+} + nB^{m-}$

t = 0 s 0 0

$t_{eq}$ s-s$\alpha$ ms$\alpha$ ns$\alpha$

$K_{sp} = [A^{n+}]^m[B^{m-}]^n$

$= [ms\alpha]^m[ns\alpha]^n$

$K_{sp} = m^m n^n (s\alpha)^{m+n}$

A very useful conclusion is derived from the **solubility product** concept. No precipitation of the electrolyte occurs if the **ionic product** is less than the **solubility product**, i.e., the solution has not reached the saturation stage.

**Case I:** When $K_{ip}<K_{sp}$, then solution is unsaturated in which more solute can be dissolved.

**Case II:** When $K_{ip} = K_{sp}$, then solution is saturated in which no more solute can be dissolved.

**Case III:** When $K_{ip} > K_{sp}$, then solution is supersaturated and precipitation takes place.

When the **ionic product** exceeds the **solubility product**, the equilibrium shifts towards left hand side, i.e., increasing the concentration of undissociated molecules of the electrolyte. As the solvent can hold a fixed amount of electrolyte at a definite temperature, the excess of the electrolyte is thrown out from the solution as precipitate.

Thus, for the precipitation of an electrolyte, it is necessary that the **ionic product** must exceed its **solubility product** For example, if equal volumes of 0.02 M $AgNO_3$solution and 0.02 M $K_2CrO_4$ solution are mixed, the precipitation of $Ag_2CrO_4$ occurs as the **ionic product** exceeds the **solubility product** of $Ag_2CrO_4$ which is $2 \times 10^{-12}$

In the resulting solution,

$[Ag^+] = 0.02/2 = 0.01 = 1*10^{-2}$ M

and $[CrO^{2-}_4] = 0.02/2 = 0.01 = 1*10^{-2}$ M

**Ionic product** of $Ag_2CrO_4 = [Ag^+]^2 [CrO^{2-}_4]$

$= (1\times10^{-2})^2 (1\times10^{-2})$

$= 1 \times 10^{-6}$

As $1 \times 10^{-6}$ is higher than $2 \times 10^{-12}$ so precipitation of $Ag_2CrO_4$ will occur.

## Applications of Solubility Product

### (i) Purification of Common Salt

Natural common salt consists of many insoluble and soluble impurities. Saturated solution of common salt is prepared and insoluble impurities are filtered off. Hydrogen chloride gas (HCl) is circulated through the saturated solution. HCl and NaCl dissociate into their respective ions as:

$$NaCl \leftrightarrow Na^+ + Cl^-$$

$$HCl \leftrightarrow H^+ + Cl^-$$

The concentration of $Cl^-$ ions increases considerably in solution due to ionization HCl. Hence, the ionic product $[Na^+][Cl^-]$ exceeds the solubility product of sodium chloride and, therefore, pure sodium chloride precipitates out from solution.

### (ii) Salting Out of Soap

Soap is a sodium salt of higher acids. From the solution, soap is precipitated by the addition of concentrated solution of sodium chloride. Soap and sodium chloride are present in the form of ions.

$$C_nH_{2n+1}\,COONa \leftrightarrow C_nH_{2n+1}\,COO^- + Na^+$$

Soap

$$NaCl \leftrightarrow Na^+ + Cl^-$$

Thus, the concentration of Na+ ions increases considerably on addition of NaCl solution. Hence, the ionic product $[C_nH_{2n+1}COO^-]$ $[Na^+]$ exceeds the solubility product of soap and, therefore, soap precipitates out from the solution.

### (iii) Manufacture of Sodium Bicarbonate (Baking Soda)

In Solvay's soda process. $CO_2$ gas is passed through ammonical brine to precipitate out $NaHCO_3$.

$$NH_4OH + CO_2 \rightarrow NH_4HCO_3$$

$$NH_4HCO_3 + NaCl \rightarrow NaHCO_3 + NH_4Cl$$

Here, $NaHCO_3$ will precipitate first as it has low solubility product compared to $NH_4Cl$, $NH_3HCO_3$ and NaCl. Hence baking soda ($NaHCO_3$) can be quantifiably estimated.

### (iv) Application of Solubility Product in Quantitative Analysis

#### 1. Estimation of Barium as Barium Sulphate

$H_2SO_4$ as precipitating agent is added to the aqueous solution of $BaCl_2$.

$$BaCl_2 + H_2SO_4 \rightarrow BaSO_4 + 2HCl$$

Precipitation of $BaSO_4$ takes place when its ionic product exceeds solubility product. $H_2SO_4$ is added in slight excess to ensure complete precipitation. Large excess of $H_2SO_4$ is harmful for complex formation.

**2. Estimation of Silver as Silver Chloride**

NaCl solution is added to the silver nitrate solution, slight excess of NaCl is added to ensure complete precipitation.

$$AgNO_3 + NaCl \rightarrow AgCl + NaBO_3$$

Again, precipitation of AgCl takes place when ionic product of AgCl exceeds its solubility product.

**3.** In a similar manner. We estimate lead as lead chromate, calcium as calcium oxalate, etc.

**(i) Precipitation of the Sulphides of Group II and IV**

Hydrogen sulphide is a weak electrolyte and is used for the precipitation of various sulphides of group II and IV in quantitative analysis.

It ionizes to a small extent in water:

$$H_2S \leftrightarrow 2H^+ + S^{2-}$$

Applying law of mass action,

$$K = ([H^+]^2 [S^{2-}])/[H_2 S]$$

The concentrations of S2⁻ ions can be decreased by increasing concentration of $H^+$ions and it can be increased by decreasing concentration of $H^+$ ions. In group II, lower concentration of sulphide ions is required as the solubility products of the sulphides of group II are low while higher concentration of sulphide ions is required in group IV as the solubility products of the suphides of group IV are high.

Ksp Values of Ionic Compounds

| Name | Formula | $K_{sp}$ |
|---|---|---|
| Aluminum hydroxide | $Al(OH)_3$ | $1.9 \times 10^{-33}$ |
| Barium carbonate | $BaCO_3$ | $2.6 \times 10^{-9}$ |
| Calcium carbonate | $CaCO_3$ | $5.0 \times 10^{-9}$ |
| Calcium fluoride | $CaF_2$ | $1.5 \times 10^{-10}$ |
| Lead(II) chloride | $PbCl_2$ | $1.2 \times 10^{-5}$ |
| Lead(II) chromate | $PbCrO_4$ | $2.8 \times 10^{-13}$ |
| Silver chloride | $AgCl$ | $1.8 \times 10^{-10}$ |
| Silver sulfate | $Ag_2SO_4$ | $1.2 \times 10^{-5}$ |

**Example**

The $K_{sp}$ for AgCl is 1.8e-10 $M^2$. What is the molar solubility of AgCl in pure water?

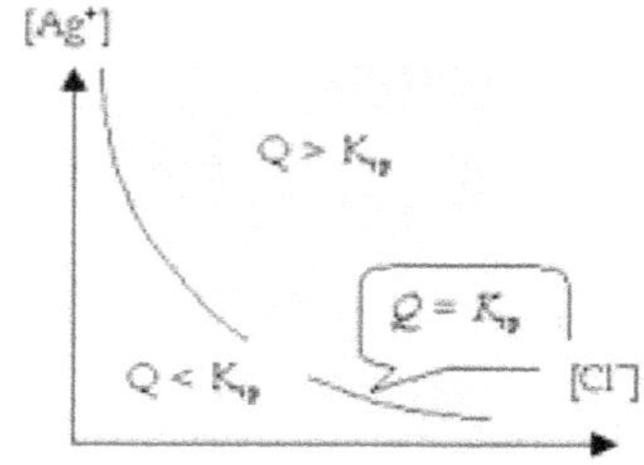


**Solution**

Let $x$ be the molar solubility, then

$$AgCl = Ag^+ + Cl^-$$

$$x \quad x$$

$$x = (1.8e\text{-}10\ M^2)^{1/2}$$

$$= 1.3e\text{-}5\ M$$

The solubility product, $K_{sp}$ is a better indicator than the usual solubility specification of g per 100 mL of solvent or moles per unit volume of solvent.

For the AgCl case, when the cation concentration is not the same as the anion concentration ($[Ag^+]$ =/= $[Cl^-]$) solubility of AgCl cannot be defined in terms of moles per L. In this case, the system can be divided into three zones. The condition $[Ag^+]\ [Cl^-] = K_{sp}$, is represented by a line which divides the plane into two zones.

When $[Ag^+]\ [Cl^-] < K_{sp}$, no precipitate will be formed.

When $[Ag^+]\ [Cl^-] > K_{sp}$, a precipitate will be formed.

When AgCl and NaCl dissolve in a solution, both salts give $Cl^-$ ions. The effect of $[Cl^-]$ on the solubility of AgCl is called the Common ion effect

**Example**

The $K_{sp}$ for $Ag_2CrO_4$ is 9e-12 $M^3$. What is the molar solubility of $Ag_2CrO_4$ in pure water?

**Solution**

Let $x$ be the molar solubility of $Ag_2CrO_4$, then

$$Ag_2CrO_4 = 2\ Ag^+ + CrO_4^{2-}$$

$$2\,x \qquad x$$

$$(2\,x)^2\ (x) = K_{sp}$$

$$x = \{(9e\text{-}12)/4\}^{(1/3)}$$

$= 1.3e^{-4}$ M

$[Ag^+]$ = 2.6e-4 M and the molar solubility is $1.3e^{-4}$ M.

**Example**

The $K_{sp}$ for $Cr(OH)_3$ is 1.2e-15 $M^4$. What is the molar solubility of $Cr(OH)_3$ in pure water?

**Solution**

Let $x$ be the molar solubility of $Cr(OH)_3$, then you have

$Cr(OH)_3 = Cr^{3+} + 3\ OH^-$

$x \quad 3\,x$

Thus,

$x\,(3\,x)^3 = 1.2e\text{-}15$

$x = \{(1.2e\text{-}15)/27\}^{(1/4)}$

$= 8.2e\text{-}5$ M

**Example**

In an experiment the molar solubility of $Bi_2S_3$ is $1.8e^{-15}$ M. Find the value of $K_{sp}$ of the compound?

**Solution**

If the molar solubility of $Bi_2S_3$ is 1.8e-15, then

$Bi_2S_3 = 2\ Bi(3+) + 3\ S^{2-}$

3.6e-15 5.4e-15

$K_{sp} = (3.6e\text{-}15)^2\,(5.4e\text{-}15)^3$

$= 2.0e\text{-}72\ M^5$

### 6.2.3 Checking of Formation of Precipitate

If 15 mL of 0.024-M lead nitrate is mixed with 30 mL of 0.030-M potassium chromate, then

$Pb(NO_3)_2$ (aq) + $K_2CrO_4$ (aq) D $PbCrO_4$ (s) + 2 $KNO_3$ (aq)

$Pb(NO_3)_2$ (aq) + $K_2CrO_4$ (aq) D $PbCrO_4$ (s) + 2 $KNO_3$ (aq)

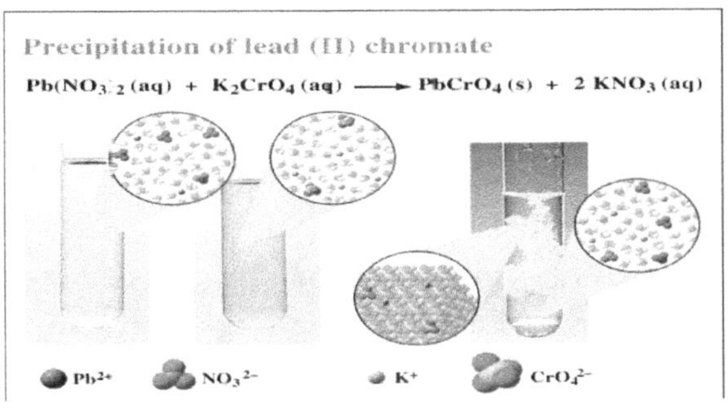

**Step 1: Checking of Sparingly Soluble Salt Formation**

In case of double replacement reaction to take place and produce $PbCrO_4$ , the salt should have very small Ksp, so that it can precipitate from the mixture. The solubility equilibrium is:

$PbCrO_4$ (s) D $Pb^{2+}$ (aq) + $CrO_4^{2-}$ (aq)

Ksp = 2 x $10^{-16}$ = $[Pb^{2+}][CrO_4^{2-}]$

If a precipitate forms, it means the solubility equilibrium has shifted back if Qsp > Ksp in mixture.

**Step 2: Find the Concentrations of the Ions that Form the Sparingly Soluble Salt**

Now mixing the two solutions so that the concentrations of $Pb^{2+}$ and $CrO_4^{2-}$ gets diluted. This involves calculation such as:

Dilution: $C_1V_1 = C_2V_2$

$[Pb^{2+}]$ =

$$\frac{C_iV_i}{V_2} = \frac{(0.024M)(15ml)}{(45mL)}$$

$$\frac{C_iV_i}{V_2} = 0.0080\,M\ pb^{2+}$$

$[CrO_4^{2-}]$ =

$$\frac{C_1V_1}{V_2} = \frac{(0.030\text{ M})(20\text{ mL})}{(45\text{ mL})}$$

$$\frac{C_1V_1}{V_2} = 0.020\text{ M }CrO_4^{2-}$$

**Step 3: Calculate Qsp for the Mixture**

Qsp = $[Pb^{2+}][CrO_4^{2-}]$

= (0.0080 M)(0.020 M)

Qsp = 1.6 x $10^{-4}$

**Step 4: Comparing Qsp to Ksp**

As Qsp >> Ksp, then a precipitate will form when two solutions are mixed. Further if Qsp = Ksp, then the mixture gets saturated while when Qsp < Ksp, then the solution will be unsaturated, so in both the cases, no precipitate will form.

## 6.3 Factors Affecting Solubility

### Common Ion Effect

The pH of a solution has good influence of solubility. Le Chatlier's principle does a great job of explaining the solubility in terms of the common ion effect and pH effects, but in some instances some puzzling results are obtained. For instance, when concentrated NH3 is added to a saturated solution of $Zn(OH)_2$ its solubility increases.

### Nature of Solute and Solvent

Solubility depends on nature of solute and solvent. The component present in ample is solvent while the component which is present in small proportion is solute. The solvent and solute may be from any state of matter, gaseous , liquid or solid. The adsorption of a non polar solute will be larger on a non polar adsorbent. But since there is competition between the solute and the solvent, the solvent should be polar in nature for the solute to be adsorbed preferentially. Moreover, since the difference in the polarities of the solute and the solvent determines their mutual solubilities, the solvent should be as polar as possible for a better absorption.

Amount of solute which dissolves will depend on type of solute. If only 1 gram of lead (II) chloride is dissolved in 100 grams of water at room temperature, then 200 grams of zinc chloride can be dissolved which shows that greater amount of zinc chloride can be dissolved in same amount of water as compared to lead II chloride.

### Temperature

It is observed that increase in temperature of solution increases the solubility of solid solute. More amount of sugar can be dissolved in warm water as compared to cold water as few solid solutes, are less soluble in warmer solutions. For gases, solubility decreases with rise in temperature of solution.

So it is clear that increasing of temperature increases solubility of solids and liquids which further decreases the solubility of gases. On adding a solute to solvent, kinetic energy of solvent molecules overcomes the attractive forces among solute particles.

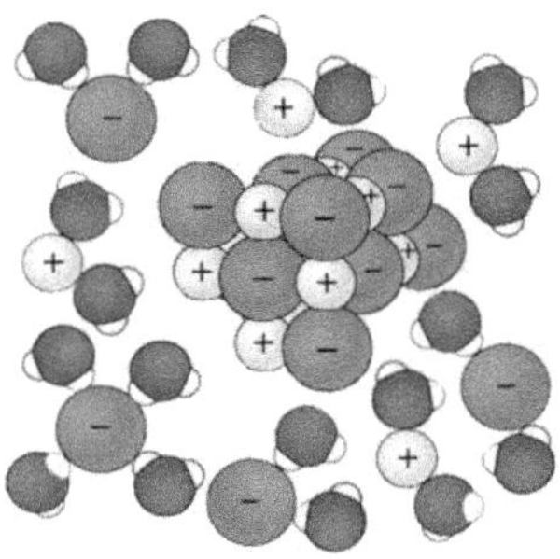

**Figure 6.9:** Mixture of Solvent.

As a result of this, solute particles leave the surface of solid and move into dissolved (aqueous) phase. As salt dissolves, positive and negative ions are pulled apart and become surrounded by water molecules.

On heating the solvent, average kinetic energies of molecules increases, so solvent is able to dislodge more particles from surface of the solute. Thus, increasing the temperature increases the solubilities of substances.

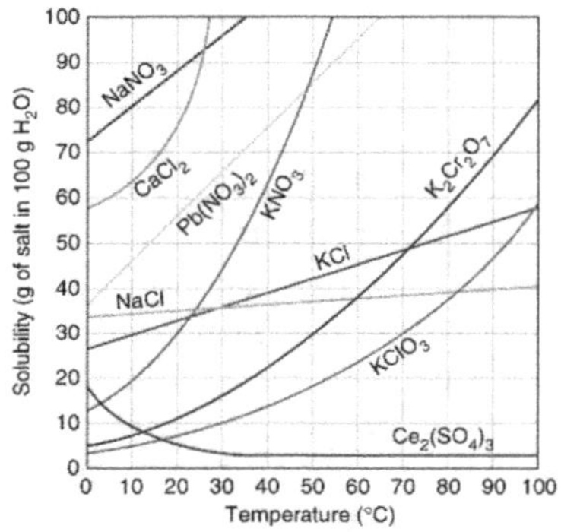


**Figure 6.10:** Solubility of $KNO_3$.

Further many solid solutes are soluble in hot water than in cold water. Figure shows that $KNO_3$ becomes much more soluble as temperature increases while NaCl shows less change in solubility.

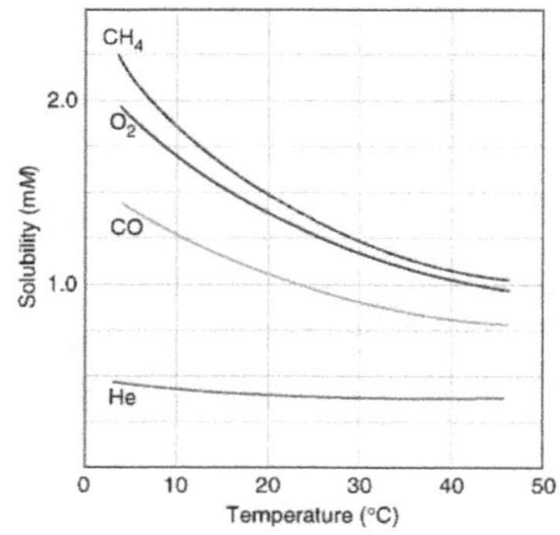


Gases dissolve in liquids similar to solids, but have simpler behavior. But, as temperature increases, solubility of a gas in liquid decreases there increasing the average kinetic energy of the gas molecules. With this, gas molecules dissolved in liquid are more likely to escape to the gas phase and not return.

**Pressure**

Change in pressure will have no noticeable effect on solubility of solid or liquid solute in a given solvent while it will affect the solubility of gaseous solute. More is the pressure, more will be the solubility of gas in a liquid.

In case of solid and liquid solutes, changes in pressure will have no effect on solubility. For gaseous solutes, an increase in pressure increases solubility and a decrease in pressure decreases solubility. When the cap on a bottle of soda pop is removed, pressure is released, and the gaseous solute bubbles out of solution. This escape of a gas from solution is called effervescence.

### Factors Affecting Rate of Solution

#### Surface Area

Although surface area does not affect the amount of solute that will dissolve, it does affect the rate of dissolution. The greater the amount of surface area, the quicker a solute will dissolve in a solvent. Thus, to make a solid solute dissolve faster we frequently powder it, thereby increasing the surface area.

#### Stirring

The rate at which a solute dissolves can also be increased by stirring the mixture. The process of stirring brings fresh solvent into contact with the solute and so permits more rapid solution.

### Strength of Solutions

#### Saturated

Suppose a small amount of salt is placed in a beaker of water. When the mixture is stirred, all the salt will dissolve. If more and more salt is added with stirring, a point will soon be reached where some of the salt settles to the bottom of the beaker. This type of solution is called a saturated solution.

#### Unsaturated

An unsaturated solution contains less of a solute than it could hold under normal conditions. Suppose that a saturated solution of glucose in a certain amount of water contains 25 g of glucose. An unsaturated glucose solution would be one that contained less than 25 g of glucose in the same amount of water.

#### Supersaturated

Under certain conditions a solvent can be made to dissolve more solute than its saturated solution can hold under the same conditions. A supersaturated solution can be prepared by adding excess solute to s saturated solution, heating that mixture, filtering off the excess solute, and then allowing the liquid to cool slowly.

#### Size of Particles

When a solute dissolves, the action takes place only at the surface of each particle. When the total surface area of the solute particles is increased, the solute dissolves more rapidly. Breaking a solute into smaller pieces increases its surface area and increases its rate of solution.

## 6.4 Features of Solid or Liquid Solutes

In a homogeneous mixture, there are two or more substances where a mixture act as solute that is dissolved in other substance which is known as solvent. The concentration of solute in a solution is a measure of quantity of solute dissolved in solvent with regards to quantity of solvent present.

It is noted that a solution is a homogeneous mixture of two or more substances in which the particles of solute in a solution is not visible with naked eye. Further a solution does not allow beam of light to scatter and is stable. Solute obtained from solution cannot be separated by filtration and is composed of single phase. Solute can be gases, liquid or solid.

**Liquid**

In case of liquid as solvent every liquids and solids can be dissolved. The examples include:

- Gas in liquid
- Oxygen in water
- Carbon dioxide in water

Liquid in Liquid

- Mixing of two or more substances of similar nature with different concentrations forming a constant such as alcoholic beverages are solutions of ethanol in water.

Solid in Liquid

- Mixing of two or more substances such as sucrose in water or sodium chloride in water.

**Solid**

- In case of solvent is solid, then gases, liquids and solids can be dissolved.

Gas in Solids

- Hydrogen dissolves well in metals

Liquid in Solid

- Mercury in gold that forms amalgam
- Water in solid salt or sugar resulting in moist solids
- Hexane in paraffin wax

Solid in Solid

- Steel, which is a solution of carbon atoms in crystalline matrix of iron atoms
- Alloys similar as bronze and others
- Polymers having plasticizers

# 7 Complex Organic Compounds

## 7.1 Organic Compounds

### 7.1.1 Introduction

The chemical compounds of living things are known as organic compounds because of their association with organisms and because they are carbon-containing compounds. Organic compounds, which are the compounds associated with life processes, are the subject matter of organic chemistry. Organic compounds are called "organic" because they are associated with living organisms. These molecules form the basis for life. They are studied in great detail in the chemistry disciplines of organic chemistry and biochemistry. Among the numerous types of organic compounds, four major categories are found in all living things: carbohydrates, lipids, proteins, and nucleic.

Organic compounds are the basis of the carbon cycle of Earth. The carbon cycle includes various biogeochemical processes that transfer carbon between living and non-living systems. The carbon cycle has existed for billions of years and is crucial to the continuity of life as we know it. That means that carbon is a very important element and organic compounds are very important compounds. But why exactly are they so important? The main reason for the value of carbon is that it can form four bonds at once, allowing it to form complex, flexible molecules.

### 7.1.2 Types of Organic Compounds

Simple organic compounds are usually the chemical byproducts of life. Many of these simple organic compounds react strongly with oxygen, causing them to burn rapidly and at a high temperature. Most are even combustible.

There are four main types or classes of organic compounds that are found in all living things like:

- Carbohydrates,
- Lipids and fats
- Proteins,
- Nucleic acids

Carbohydrates are made from carbon, hydrogen and oxygen which are organic molecules typically classified according to their structure. Carbohydrates can be found in many of the foods that we eat such as on the rice, on the potatoes, corns, avocado and so much more. This is one of the sources of energy and heat and also serves as fuel to keep our body going.

Carbohydrates are primary energy source for humans. Chemically, carbohydrates are organic molecules where carbon, hydrogen and oxygen bond together in ratio $C_x(H_2O)_y$, where $x$ and $y$ are numbers which varies as per specific carbohydrates. Animals break down carbohydrates at the time of metabolism in order to release energy. For example, the chemical metabolism of the sugar glucose is shown below:

$$C_6H_{12}O_6 + 6\ O_2 \rightarrow 6\ CO_2 + 6\ H_2O + \textit{energy}$$

Animals get carbohydrates by eating foods such as potatoes, rice, breads, etc. These carbohydrates are prepared by plants during photosynthesis where plants harvest energy from sunlight in order to run reverse reaction as shown:

$$6\ CO_2 + 6\ H_2O + \underset{\textit{(from sunlight)}}{\text{energy}} \rightarrow C_6H_{12}O_6 + 6\ O_2$$

As potato is primary chemical storage system containing glucose molecules prepared during photosynthesis, so in potato, the glucose molecules are bound together in long chain. As it turns out, there are two types of carbohydrates, simple sugars and those carbohydrates that are made of long chains of sugars - the complex carbohydrates; hence carbohydrates are of two types:

- Simple
- Complex

**Simple Sugars**

Carbohydrates are made from carbon, hydrogen and oxygen. These elements form starch and sugar where sugar is considered as most abundant carbohydrates classified as:

- Monosaccharide
- Disaccharide
- Polysaccharide

As carbohydrates are made up of saccharide units, so carbohydrates that has only one sugar unit is known as monosaccharides, two sugar units are known as disaccharides which are referred to as simple sugars.

Simple sugars are sweet in taste and are broken down quickly in body to release energy. Among all, two most common monosaccharides are glucose and fructose. Glucose is main form of sugar which is stored in human body for energy while Fructose is main sugar found in fruits. Both glucose and fructose have similar chemical formula ($C_6H_{12}O_6$), but have different structures as shown:

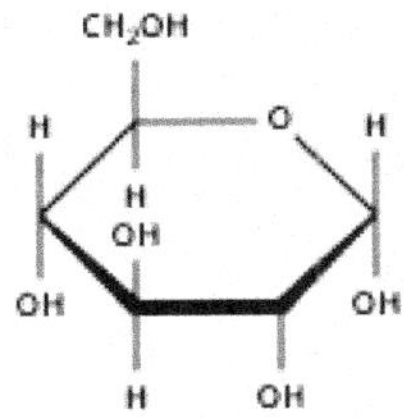


**Figure 7.1**: Glucose

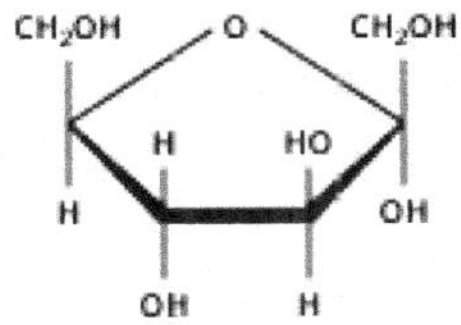


**Figure 7.2**: Fructose

Disaccharides have two sugar units bonded together. For example, common table sugar is sucrose, a disaccharide that consists of a glucose unit bonded to a fructose unit:

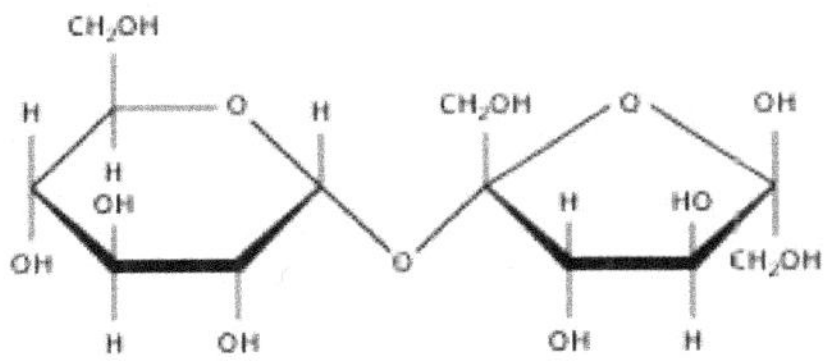


**Figure 7.3**: Sucrose

The disaccharides are double-sugars like the sucrose and the lactose.

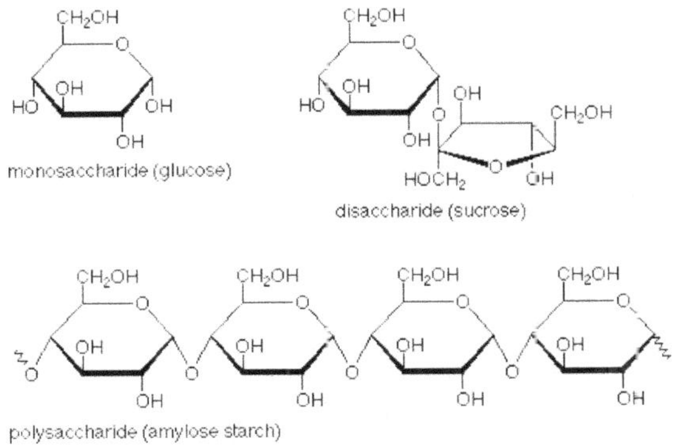


**Figure 7.4**: Types of Carbohydrates.

## Complex Carbohydrates

Complex carbohydrates are polymers of simple sugars having long chains of simple sugar units bonded together. In case of potato, there are complex carbohydrate starch which serves as a polymer of monosaccharide glucose. Complex carbohydrates are called polysaccharides since they have more than two sugar groups linked together. The polysaccharides are the multiple sugars like the glycogen.

Starch is main polysaccharide that are used by plants to storing glucose for later use as energy. Plants normally store starch in seeds or organs like rice, beans, wheat, corn, potatoes, etc. Consumption of starch by humans allows occurrence of enzyme in saliva and in intestines for breaking of bonds among repeating glucose units, thereby allowing sugar to absorbed in blood.

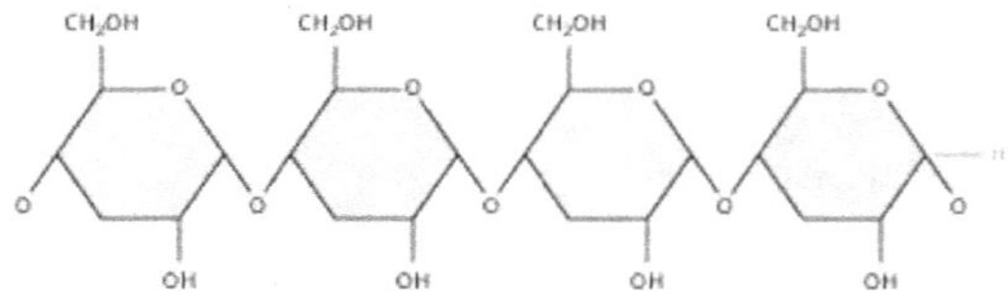


**Figure 7.5**: Starch Molecule.

Glycogen, which is a polymer of glucose is polysaccharide which are used by animals for storing their energy. Extra glucose is bonded together forming glycogen molecules, which animal stores in liver and muscle tissue as a source of energy. Starch and glycogen are polymers of glucose, while starch is long and straight chain of glucose units whereas glycogen has branches of glucose units as shown in figure below:

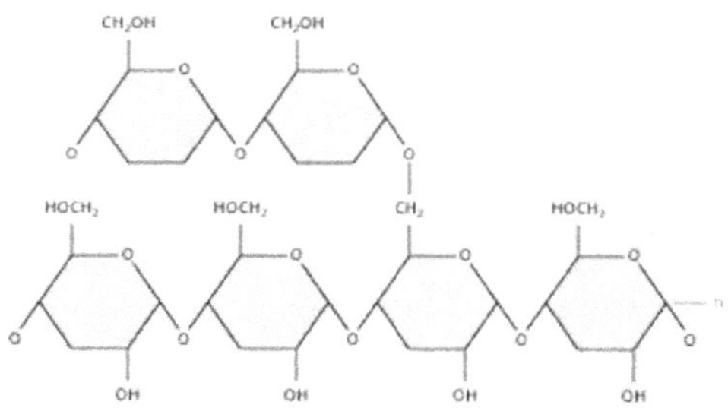


**Figure 7.6**: Glycogen Molecule.

Cellulose which is another polysaccharide is a third polymer of monosaccharide glucose. It is different from starch and glycogen as in this, glucose units forms two-dimensional structure, with hydrogen bonding that holds nearby polymers together giving molecule added stability.

Cellulose, also known as plant fibers that are not digested by humans and passes through digestive tract without absorbing in body. Different animals like cows and termites have bacteria in digestive tract which supports in digestion. Being an hard materials, in plants cellulose is used as structural molecule which gives support to leaves, stem and plant parts.

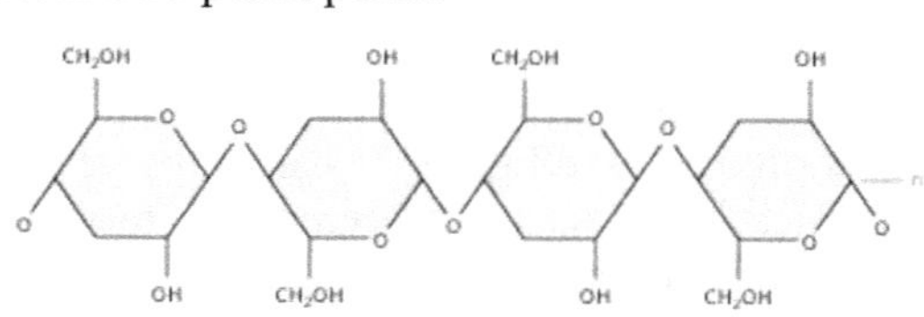


**Figure 7.7**: Cellulose Molecule.

## Fats and Lipids

Fats and lipids are made from elements of carbon, hydrogen and oxygen with lesser oxygen. Lipids are made of carbon, hydrogen, and oxygen atoms. Lipids have high hydrogen to oxygen ratio as compared to with carbohydrates and are used for storing energy for building structures and as signal molecules for cells communication with each other.

The fats and lipids fatty acids and glycerol are known as glycerin are obtained from sources like pork, coconut and peanuts. These organic compounds are very much convenient in storing the reserve energy in the body. Since the lipids are organic molecules of carbon, hydrogen and oxygen with high ratio of hydrogen to oxygen atoms includes steroids which is composed of many hormones, waxes and fats.

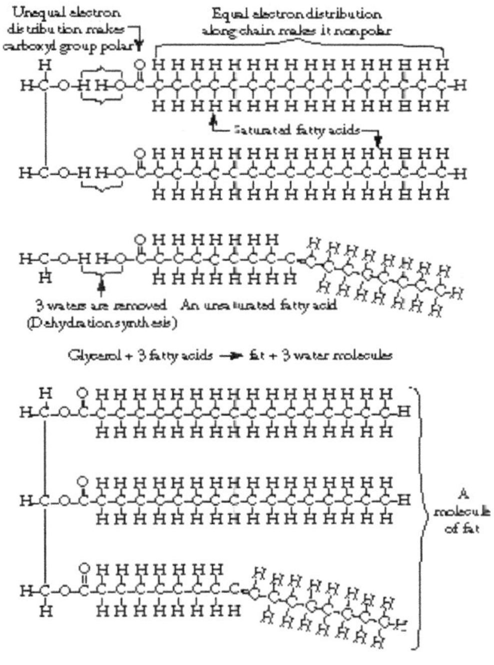


**Figure 7.8**: Fat Molecule from Glycerol Molecule with 3 Fatty Acid Molecules.

Fat molecules contains glycerol molecule and 1, 2 or 3 molecules of fatty acids. Glycerol molecule has 3 hydroxyl (–OH) groups. Fatty acid is a long chain of carbon atoms with carboxyl (–COOH) group at one end which in fat are similar or different. They are bound to the glycerol molecule by a process that involves the removal of water.

Certain fatty acids have one or more double bonds in their molecules. Fats that include these molecules are unsaturated fats. Other fatty acids have no double bonds. Fats that include these fatty acids are saturated fats. In most human health situations, the consumption of unsaturated fats is preferred to the consumption of saturated fats.

Fats stored in cells usually form clear oil droplets called globules because fats do not dissolve in water. Plants often store fats in their seeds, and animals store fats in large, clear globules in the cells of adipose tissue. The fats in adipose tissue contain much concentrated energy. Hence, they serve as a reserve energy supply to the organism. The enzyme lipase breaks down fats into fatty acids and glycerol in the human digestive system.

**Lipid Examples**

- Cholesterol
- Paraffin
- Olive oil
- Margarine
- Cortisol
- Estrogen
- Phospholipid bilayer

## Proteins

Proteins are organic compounds made from carbon, hydrogen, oxygen, nitrogen, sulfur and phosphorus.

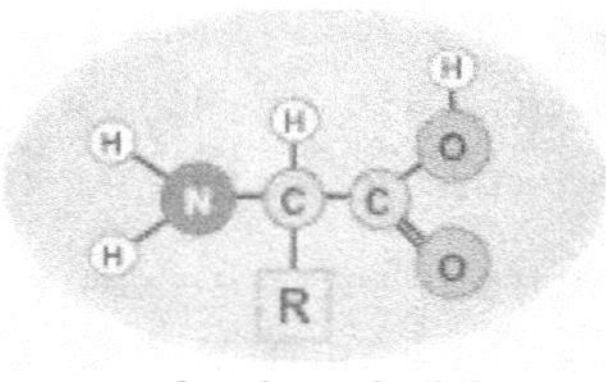


**Figure 7.9**: Amino Acids.

These are kinds of large heavy and generally complex molecules of amino acids. A protein may be made from a single polypeptide chain or from complex structure where polypeptide subunits forming a unit.

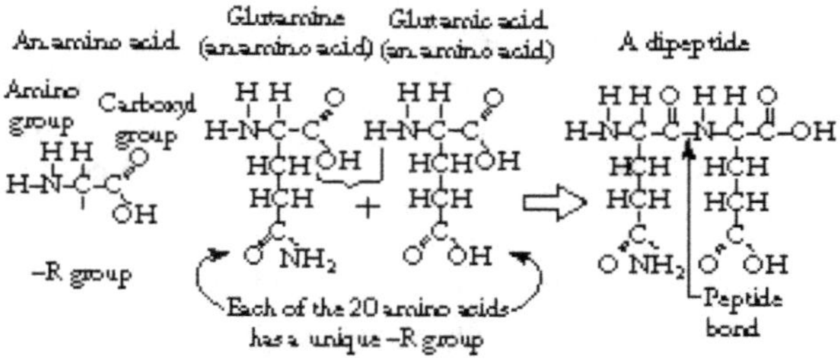


**Figure 7.10**: Structure of Amino Acids.

Proteins consist of hydrogen, oxygen, carbon and nitrogen atoms. Some proteins contain other atoms like sulfur, phosphorus, iron, copper or magnesium. These organic compounds are important to the body because they are responsible for repairing worn-out tissues; they also act as catalyst for a number of chemical reactions inside the body as well as they are also responsible for the contraction of the muscles. It is considered that all lives from unicellular to multi-cellular are all based on proteins.

Proteins involves in many functions in cells such as building structure, catalyzing biochemical reactions, immuning response and in replicating genetic material. The examples of protein are:

- Enzymes
- Collagen
- Keratin
- Albumin
- Hemoglobin
- Myoglobin
- Fibrin

## Nucleic Acids

Nucleic acids are organic compounds having carbon, hydrogen, nitrogen and phosphorus. They have larger molecules and are composed of small units known as nucleotides. Each nucleotide has carbohydrate molecule, phosphate group, and nitrogen containing molecule due to their properties as nitrogenous base. In nucleic acid there are two monocyclic bases as pyrimidines, and two bicyclic bases as purines each having one N-H site at which organic substituent is attached.

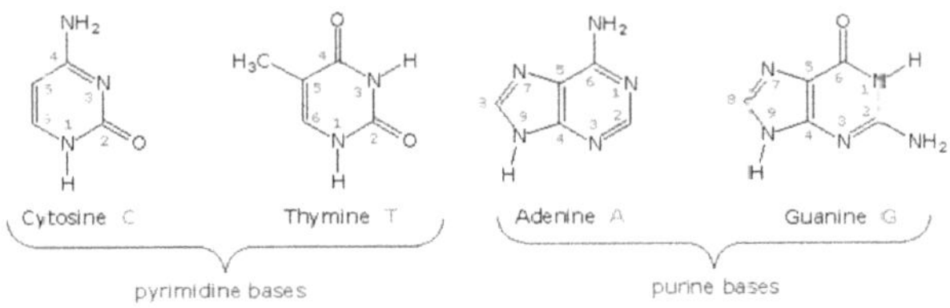


**Figure 7.11:** Purine and Pyrimidine Bases.

As purine and pyrimidine are base components of nucleotides, it shows that, three of them exist as hydroxy pyrimidine or purine tautomers with aromatic heterocyclic ring. Inspite of extra stabilization of aromatic ring, compounds adopt amide like structures as shown in Figure 7.12.

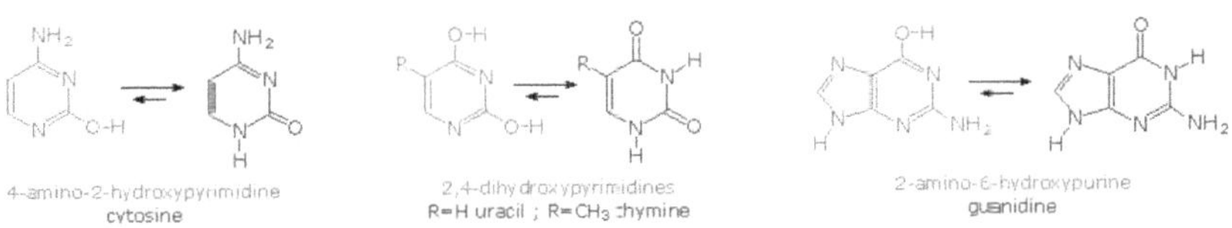


**Figure 7.12**

Nucleic acid has two classes;

- RNA or ribonucleic acid
- DNA or deoxyribonucleic acid

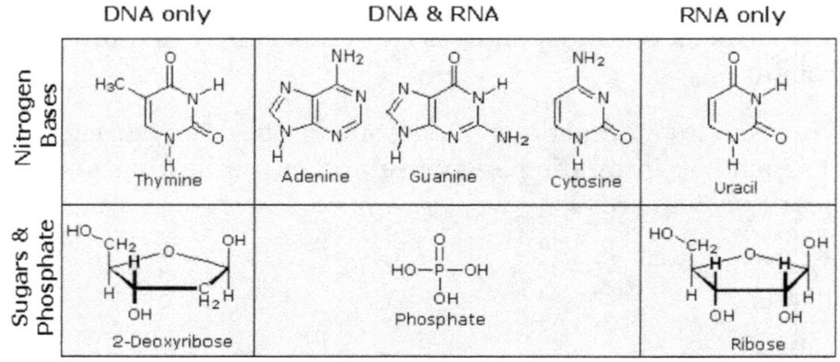


**Figure 7.13**: Components.

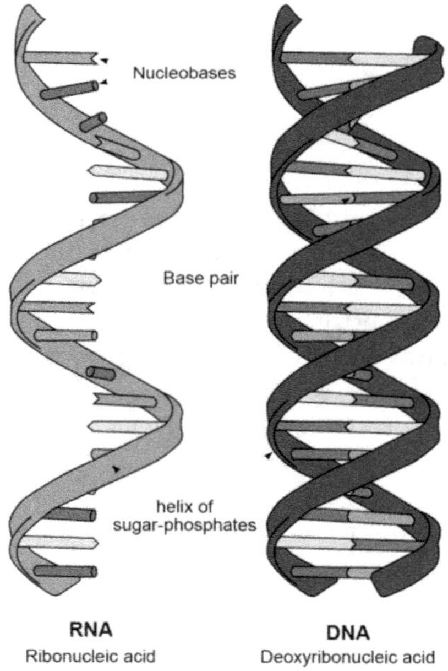


**Figure 7.14**

## Deoxyribonucleic Acid (DNA)

DNA is cellular molecule that contains carbohydrate deoxyribose instructions for doing cell functions and is found mainly in nucleus of the cell. When a cell divides, its DNA gets copied and passed from one cell generation to other. It is organized in chromosomes and contains programmatic instructions for cellular activities. During the organisms produce their offspring, such instructions are passed through DNA. It exists as double stranded molecule with twisted double helix shape.

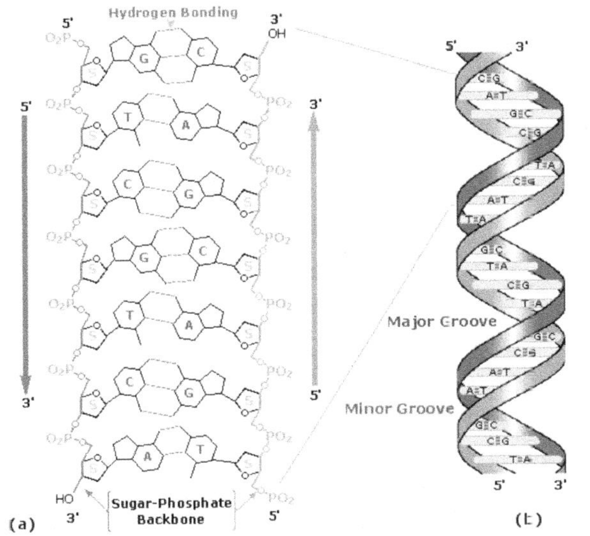


**Figure 7.15**: Instructions and Helical Structure of DNA.

From the figure shown below, DNA of cell is tightly packed in chromosomes where initially DNA gets wrapped around with small proteins called as histones which are bead like structures that are arranged and folded in chromatin aggregates resulting in chromosomes. It is observed that a series of unfolding events takes place before the information gets encoded in DNA which are used for replication.

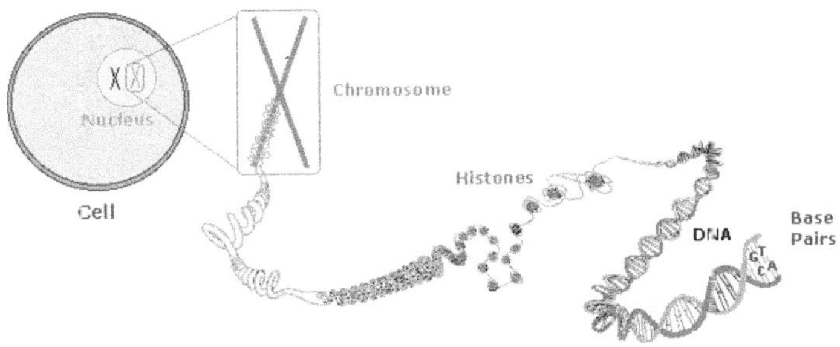


**Figure 7.16**: DNA Replication

DNA contains phosphate-deoxyribose sugar backbone and four nitrogenous bases:

- Adenine (A)
- Guanine (G)
- Cytosine (C)
- Thymine (T)

In double stranded DNA, adenine pairs with thymine (A-T) and guanine pairs with cytosine (G-C).

### Ribonucleic Acid (RNA)

RNA is essential for synthesis of proteins. The RNA serves as the chemical messenger between the DNA of the nucleus and the metabolic machinery in the cytoplasm. RNA is found in both the nucleus and the cytoplasm, a semiliquid substance that composes the volume of the cell. RNA commonly exists as single stranded molecule and is composed of phosphate ribose sugar backbone and nitrogenous bases adenine, guanine, cytosine and uracil.

Just like DNA, RNA is single-stranded which has ribose sugars instead of deoxyribose sugars that make RNA more unstable and prone to degradation. In this, the information in genetic code is passed from DNA to RNA to resulting proteins. RNA is synthesized from DNA by enzyme known as RNA polymerase at the time of transcription. The new RNA sequences are complementary to DNA template, instead of being identical copies of template which gets translated into proteins by structures known as ribosomes.

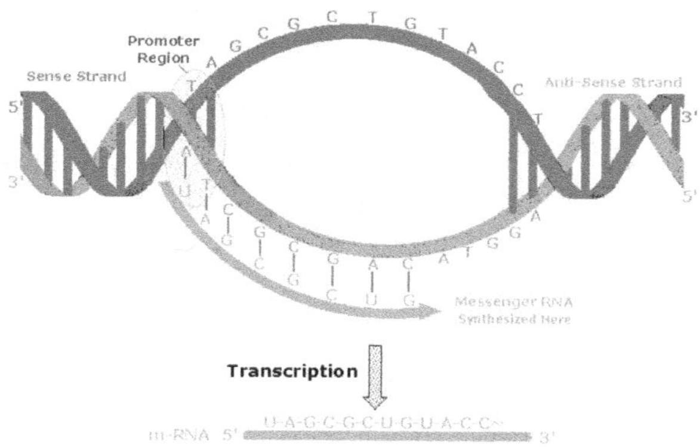


**Figure 7.17**: RNA Transcription Process.

There are three types of RNA involved in translation process:

- Messenger RNA (mRNA)
- Transfer RNA (tRNA)
- Ribosomal RNA (rRNA)

Among this, messenger RNA (mRNA) is RNA transcript or RNA copy of DNA message which is obtained at the time of DNA transcription. The messenger RNA gets translated to form proteins. Transfer RNA (tRNA) has three dimensional shapes which is required for translation of mRNA in protein synthesis. Ribosomal RNA (rRNA) ribosome's involved in protein synthesis.

## 7.1.3 Classification of Organic Compounds

### Acyclic or Open-Chain Compounds

Open chain compound or acyclic compound are compounds having linear structure without any side chains and are commonly known as straight-chain compound. Molecules of organic chemistry like alkanes and alkenes have linear and ring isomers, acyclic and cyclic, with latter classified as aromatic.

- $CH_3–CH_2–CH_2–CH_3$
- $CH_3–CH_2–CH–CH_3$

  |

  Cl

In case of molecules with 4 or more carbons, linear forms straight-chain or branched-chain isomers. The lowercase prefix n- shows straight-chain isomer like n-butane is straight-chain butane. Cycloalkanes are isomers of alkenes and not of alkanes as ring's closure will be C=C bond. Without any rings, all open-chain compounds are aliphatic. Open chain compounds have linked carbon atoms with linear or branched fashion like molecule with open chain structure. Such compounds are aliphatic compounds such as propane.

$CH_3$ - $CH_3$ - $CH_3$ propane

**Cyclic or Closed-chain Compounds**

These are hydrocarbons having ring of carbon atoms known as cyclic ring compounds.

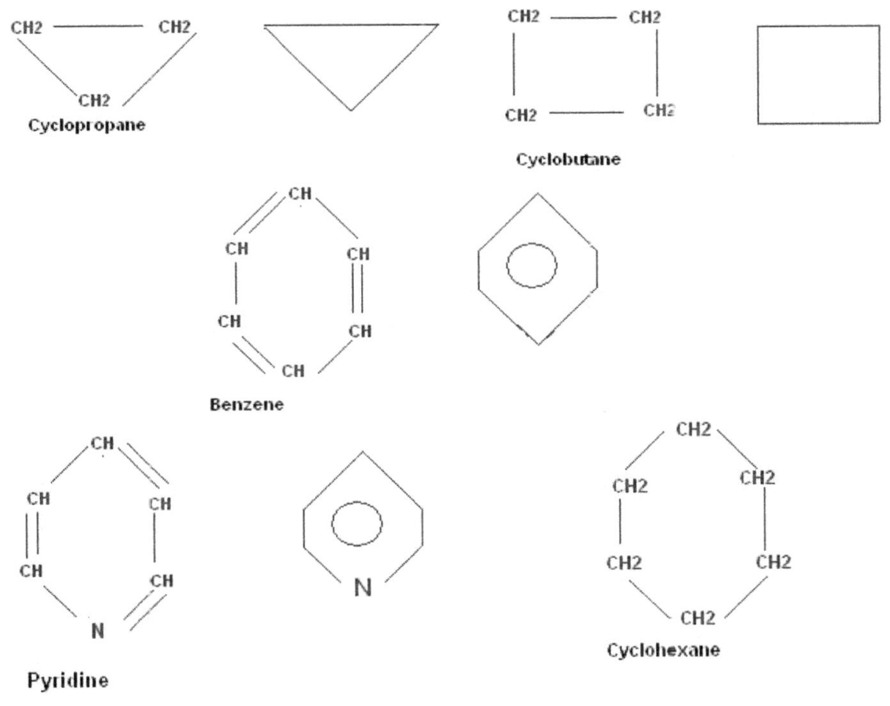


**Figure 7.18**: Cyclic Ring Compounds.

Organic compounds having closed chain of atoms are closed chain compounds or ring compounds which are classified as:

a) Homocyclic compounds

b) Heterocyclic compounds

**Homocyclic Compounds**

The cyclic organic compounds which are made of carbon atoms are homocyclic compounds or homocyclic hydrocarbons such as Cyclopentane, cyclohexane etc.

**(a) Alicyclic Compounds**

Alicyclic compounds are cyclic hydrocarbons having no alternate C-C and C=C but carries similar features of aliphatic compounds.

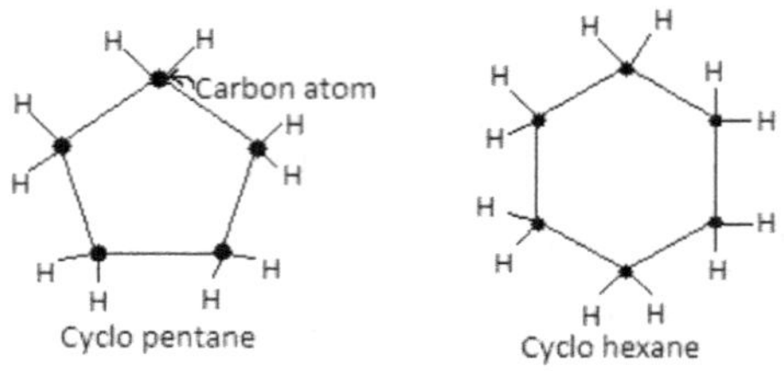


**Figure 7.20**: Alicyclic Compounds.

**(b) Aromatic Compounds**

Aromatic hydrocarbons are cyclic hydrocarbons having alternate C-C and C=C which are similar to benzene ring as benzene have a ring with six carbon atoms.

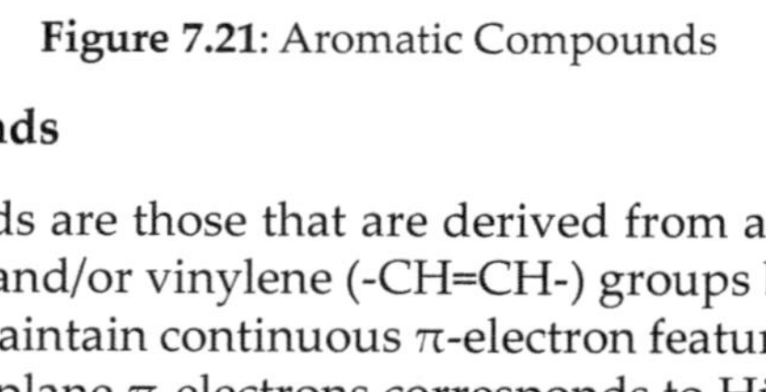


**Figure 7.21**: Aromatic Compounds

**Heterocyclic Compounds**

Heterocyclic compounds are those that are derived from arenes by replacing one or more methine (-C=) and/or vinylene (-CH=CH-) groups by trivalent or divalent hetero atoms so as to maintain continuous $\pi$-electron features of aromatic systems with number of out-of-plane $\pi$-electrons corresponds to Hückel rule (4n + 2).

**Figure 7.22**

These cyclic hydrocarbons are made of carbon atoms with other elements as N, O, S etc.

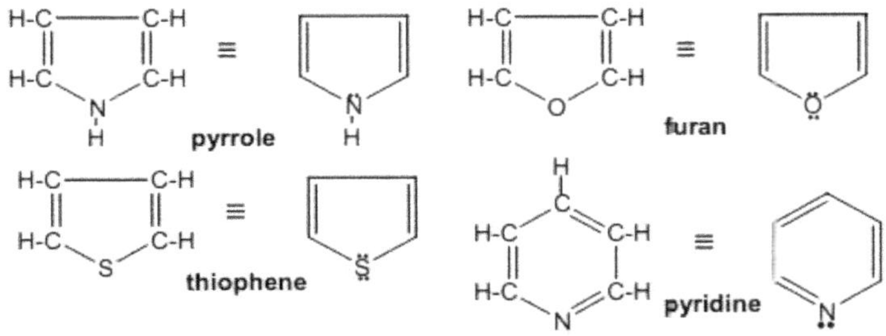


**Figure 7.23**: Heterocyclic Compounds.

## 7.2 Synthesis of Complex Organic Compounds

Organic Compound synthesis involves in dealing with formation of organic compounds. There are two main areas of research in general area of organic synthesis which includes total synthesis and methodology. It is noted that chemical synthesis is narrow and restricted to particular kind of chemical reaction which serves as direct combination reaction where two or more reactants combine to form single product such as:

$$A + B \rightarrow AB$$

where

A and B are elements or compounds

AB is compound of A and B

Combination chemical reactions include:

$2Na + Cl_2 \rightarrow 2\ NaCl$ (formation of table salt)

$S + O_2 \rightarrow SO_2$ (formation of sulfur dioxide)

$4\ Fe + 3\ O_2 \rightarrow 2\ Fe_2O_3$ (iron rusting)

$CO_2 + H_2O \rightarrow H_2CO_3$ (carbon dioxide dissolving and reacting with water forming carbonic acid)

Total synthesis appear as complete chemical synthesis of complex organic molecules from simple that are commercially available. Linear synthesis has series of steps that are done one after another till molecule is made which is often adequate for simple structure. Chemical compounds made in every step are referred as synthetic intermediates. In case of more complex molecules, convergent synthesis is preferred as pieces of final product are synthesized separately and further coupled together near the end of synthesis. Retrosynthesis involves thinking backwards from relatively complex molecules to simpler ones the disconnection approach.

Synthesising Organic Compounds is done for:

- Proof of structure of natural compound by keeping it together from simpler molecules
- Preparing compounds useful for mankind
- Preparing particular compounds to study reaction mechanisms
- Preparing of new problems and new solutions for development of reagents etc.
- In synthesis, every step involves chemical reaction, reagents and conditions for every reactions which is required to design good yield and product doing less work.

### 7.2.1 Synthesis of Substituted Benzene Derivatives

Reactions are direct as in case of $S_EAr$ and involves simple synthesis due to nature of starting materials, hence many reactions correspond to following disconnection:

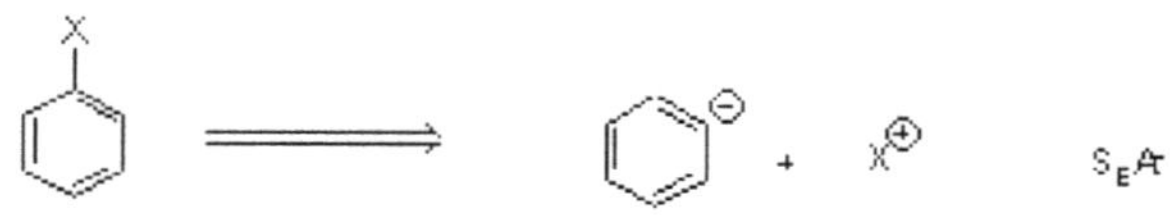


Example

1[st] decision which bond to disconnect first!

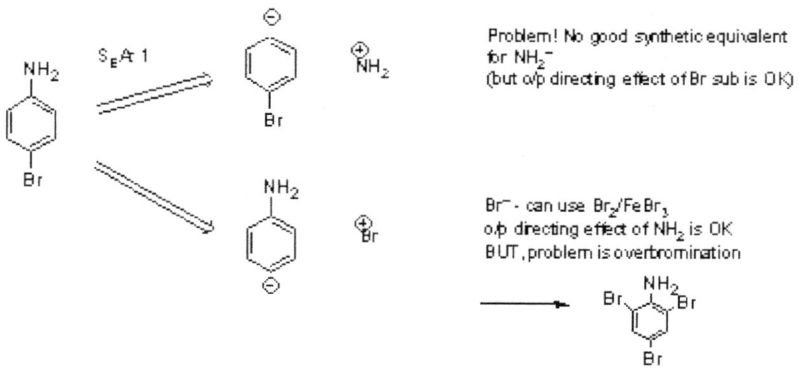

Doing monobromination on N-acyl derivative of amine shows:

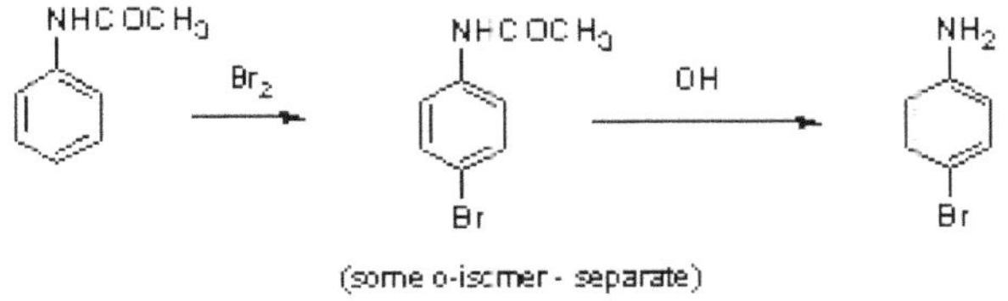


involves removing protecting group ($HO^-/H_2O$) to give required product, hence,:

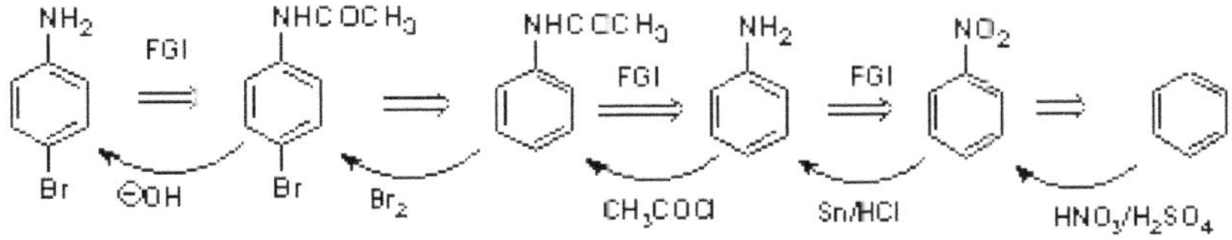

Knowing an alternative route,

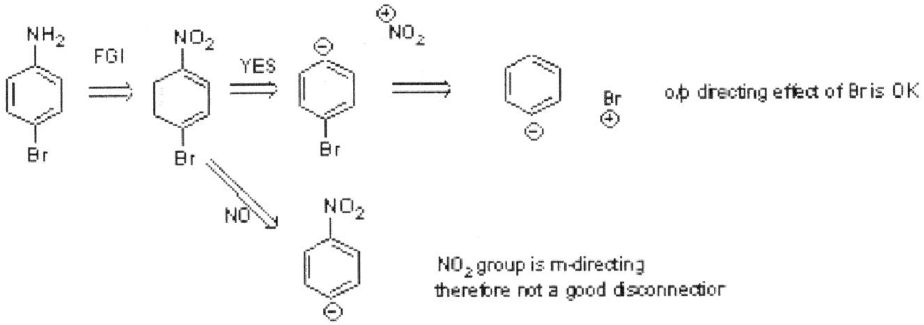


**Example**

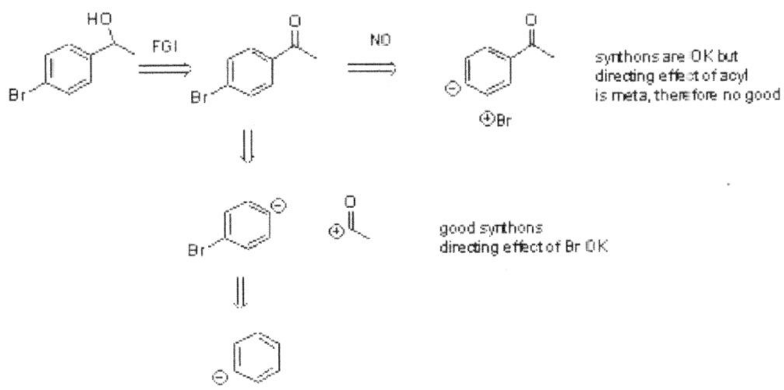


Synthesis

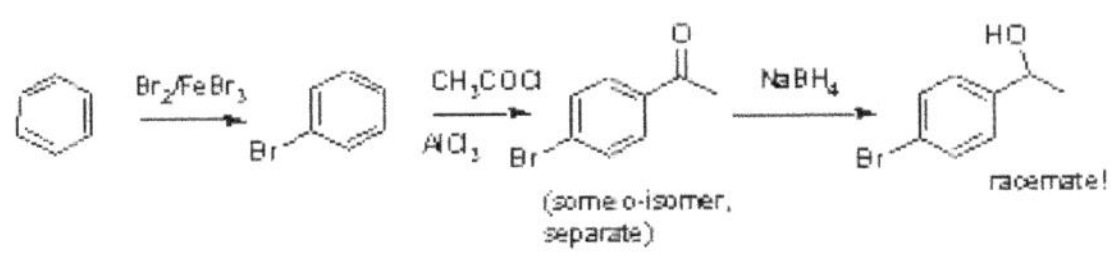

Example

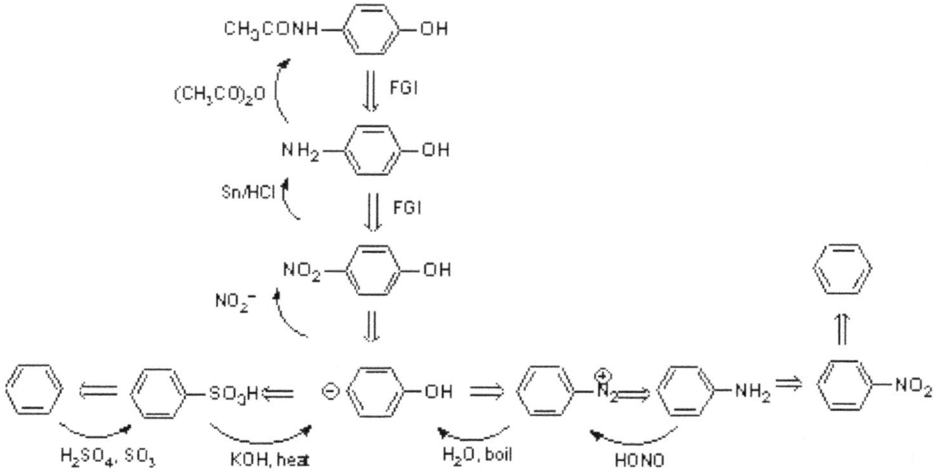

### 7.2.2 Guidelines for Designing a Synthesis

- Using retrosynthetic analysis to work backwards from TM to precursors and to RASM
- Locating functional groups in TM for most functional groups
- Examining all possible disconnections by checking which are chemically sound
- If making no progress, use FGI: ($NO_2/NH_2$; $CH_3/COOH$; C-Br/C-OH; $CHO/CH_2OH$ etc.)
- After receiving precursors to TM, repeat process on intermediates

### 7.2.3 Examples of Retrosynthetic Analysis

1

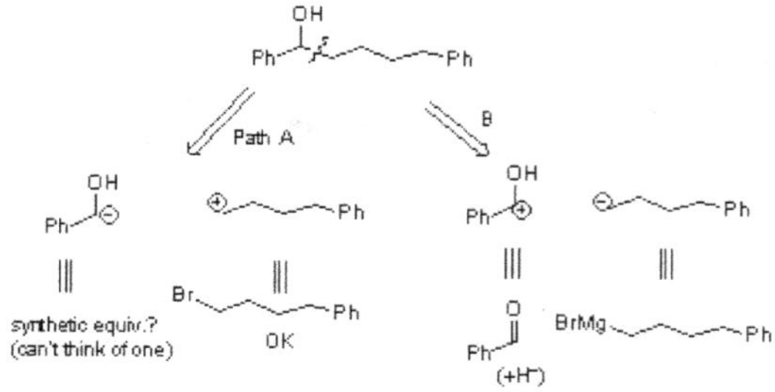

Synthesis

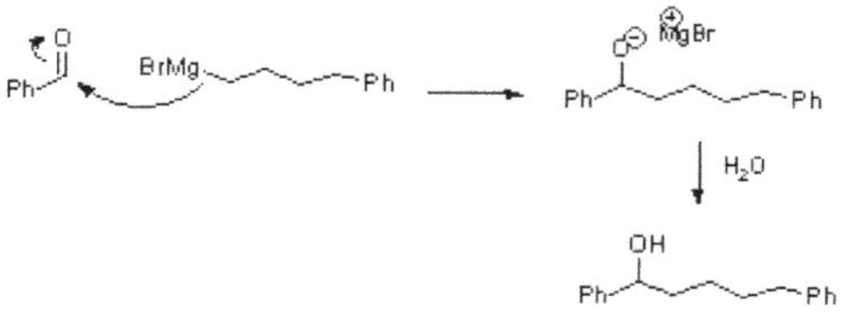

2

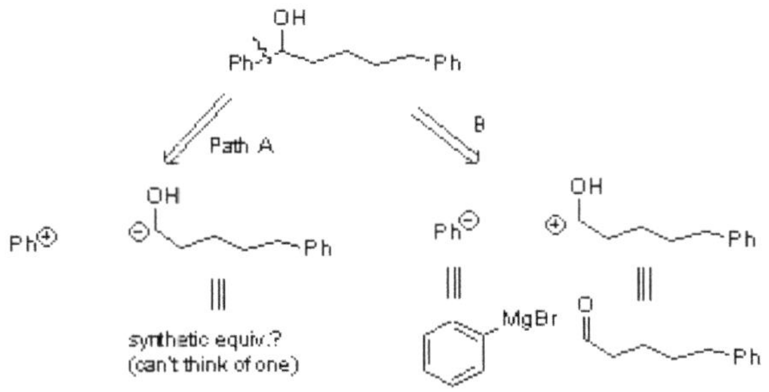


Synthesis

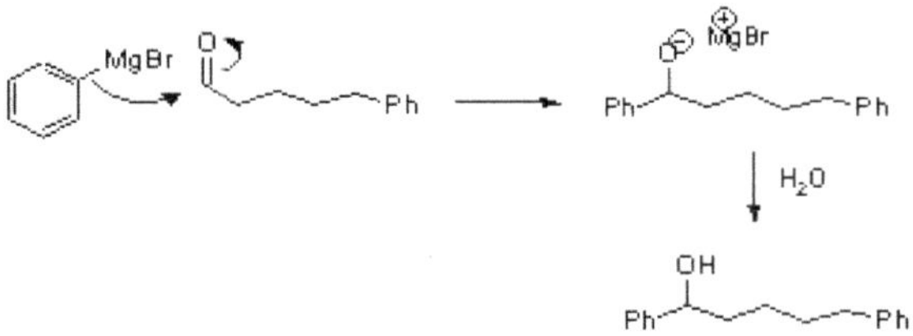

3

Synthesis

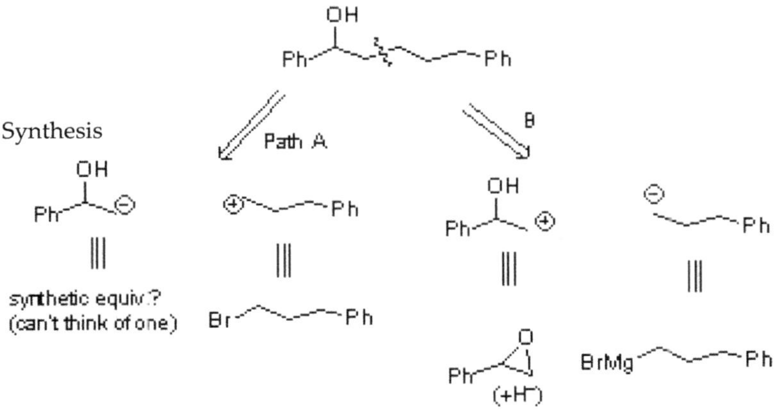


Synthesis

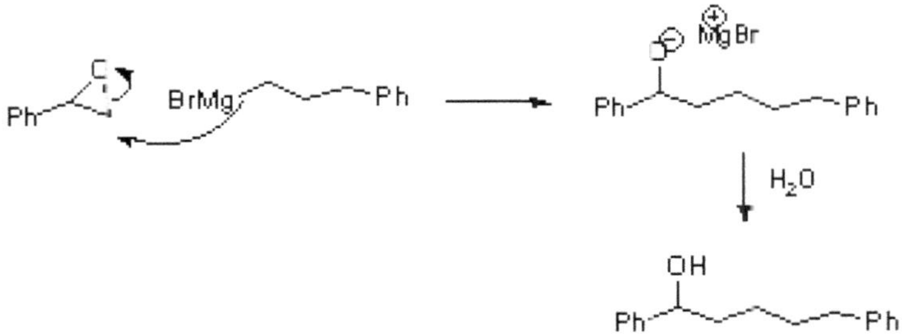

4

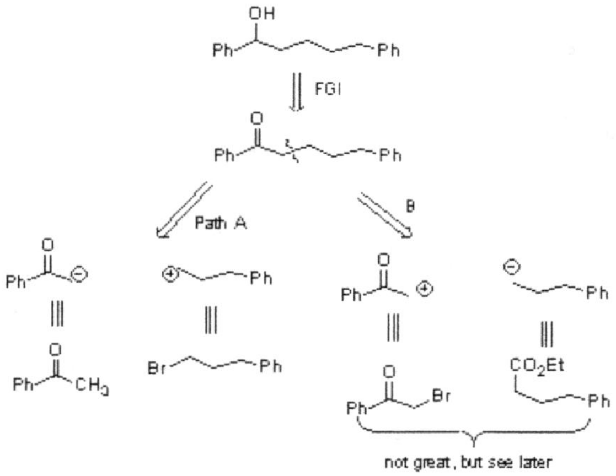


Synthesis

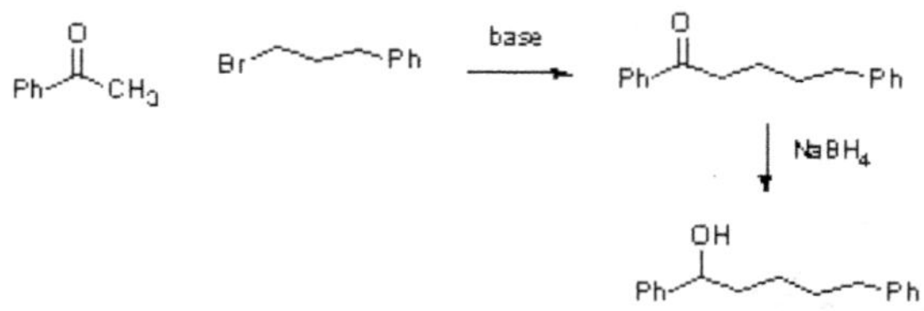


5

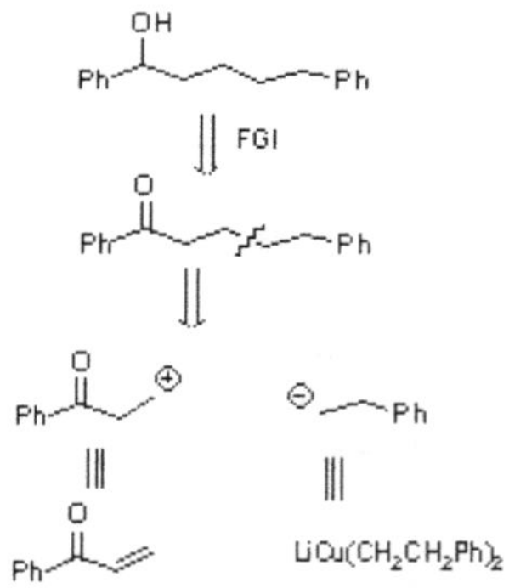


Synthesis

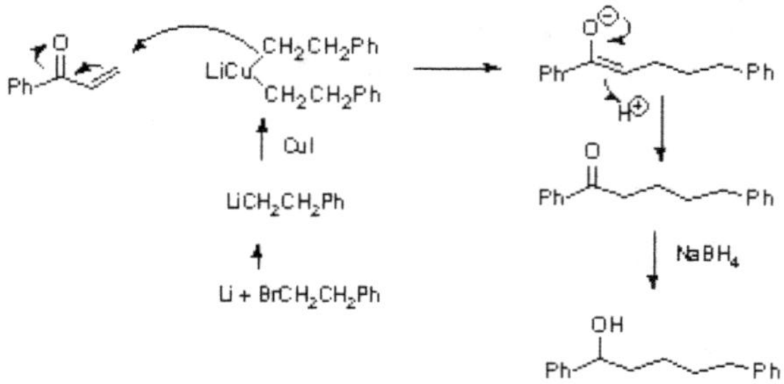

# 8 Oxidation-Reduction Equilibria

## 8.1 Oxidation States of Atoms and Ions

The oxidation state, often called the oxidation number, is an indicator of the degree of oxidation (loss of electrons) of an atom in a chemical compound. Conceptually, the oxidation state, which may be positive, negative or zero, is the hypothetical charge that an atom would have if all bonds to atoms of different elements were 100% ionic, with no covalent component. This is never exactly true for real bonds. High oxidation numbers generally correlate with high valences, although the concepts are not equivalent.

The term "oxidation" was first used by Antoine Lavoisier to signify reaction of a substance with oxygen. Much later, it was realized that the substance, upon being oxidized, loses electrons, and the use of the term "oxidation" was extended to include other reactions in which electrons are lost.

The increase in oxidation state of an atom, through a chemical reaction, is known as an oxidation; a decrease in oxidation state is known as a reduction. Such reactions involve the formal transfer of electrons: a net gain in electrons being a reduction, and a net loss of electrons being an oxidation. For pure elements, the oxidation state is zero.

Oxidation state" is defined as the charge an atom might be imagined to have when electrons are counted according to an agreed-upon set of rules:

- The oxidation state of a free element (uncombined element) is zero
- For a simple (monatomic) ion, the oxidation state is equal to the net charge on the ion
- Hydrogen has an oxidation state of +1 and oxygen has an oxidation state of −2 when they are present in most compounds. Exceptions to this are that hydrogen has an oxidation state of −1 in hydrides of active metals, e.g. LiH, and oxygen has an oxidation state of −1 in peroxides, e.g. $H_2O_2$.

The algebraic sum of oxidation states of all atoms in a neutral molecule must be zero, while in ions the algebraic sum of the oxidation states of the constituent atoms must be equal to the charge on the ion.

### 8.1.1 Ionic Compounds

If an ionic compound has two ions with a common element, such as ammonium nitrate, $NH_4NO_3$, it is usual to consider the oxidation states for each ion separately. According to that empirical formula, one can use the standard charges of hydrogen and oxygen to solve for the charge of nitrogen: $2x + 4(1) + 3(-2) = 0$, giving an average nitrogen oxidation state x = +1. However it is more correct to consider separately the ions $NH^+_4$ and $NO^-_3$ with nitrogen oxidation states of −3 and +5 respectively, because the two nitrogen atoms are in different covalent environments.

### 8.1.2 Oxidation State of Atoms

The oxidation state of an atom is often different from the formal charge often included in Lewis structures (when it is non-zero). The oxidation state is calculated by assuming that each chemical bond (except between identical atoms) is ionic so that both electrons are assigned to the more electronegative bonded atom. In contrast, the formal charge is calculated by assuming that each bond is covalent so that one electron is assigned to each bonded atom.

For example, in ammonium ion (NH+4) the oxidation state of nitrogen is −3, as all eight valence electrons are assigned to the nitrogen atom that is more electronegative than hydrogen. However, the formal charge is +1, calculated by assigning only four valence electrons (one per bond) to nitrogen. For comparison, the nitrogen in ammonia ($NH_3$) has oxidation state −3 also but a formal charge of zero. On protonation of ammonia to form ammonium, the formal charge on nitrogen changes, but its oxidation state does not.

**Guidelines for Assigning Oxidation States**

- The oxidation state of any element such as Fe, $H_2$, $O_2$, $P_4$, $S_8$ **is zero (0).**
- The oxidation state of oxygen in its compounds is -2, except for peroxides like $H_2O_2$, and $Na_2O_2$, in which the oxidation state for O is -1.
- The oxidation state of hydrogen is +1 in its compounds, except for metal hydrides, such as NaH, LiH, etc., in which the oxidation state for H is -1.
- The oxidation states of other elements are then assigned to make the algebraic sum of the oxidation states equal to the net charge on the molecule or ion.
- The following elements usually have the same oxidation states in their compounds:
    - +1 for alkali metals - Li, Na, K, Rb, Cs;
    - +2 for alkaline earth metals - Be, Mg, Ca, Sr, Ba;
    - -1 for halogens except when they form compounds with oxygen or one another;

Such rules shows description of special cases like $H_2O_2$ and $Na_2O_2$. The above rule deals with hydride and apart from this, the description about oxidation states for

H and O are +1 and -2 shown in a compound with oxidation of other elements that is assigned by making algebraic sum of oxidation states equals to net charge on molecule or ion.

## 8.2 Oxidation-Reduction Reactions

Oxidation-reduction reaction is a type of chemical reaction which involves transfering of electrons among two species. It is a type of chemical reaction where oxidation number of a molecule, atom or ion changes by gaining or losing an electron. Redox reactions are common and vital to some of the basic functions of life, including photosynthesis, respiration, combustion, and corrosion or rusting.

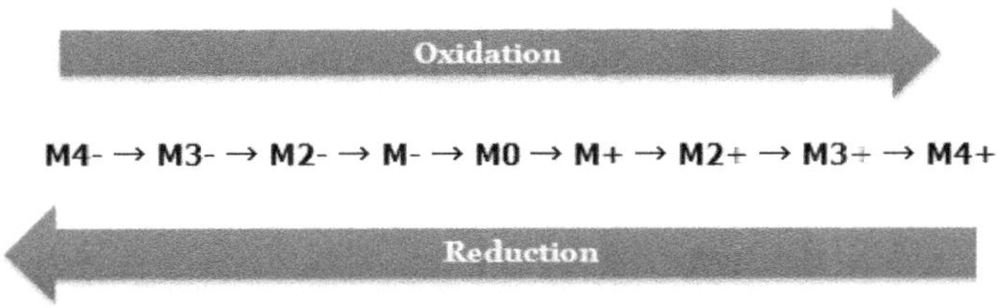


**Figure 8.1:** Oxidation-Reduction Reaction.

Redox reactions comprises reduced half and oxidized half which always occur together. In reduced half, gain of electrons and decrease in oxidation number exist while in oxidized half, loss of electrons and oxidation number increases which can be remembered as OIL RIG describing "oxidation is loss" and "reduction is gain," and LEO GER describes "loss of e- = oxidation" and "gain of e- = reduced." There is no net change in the number of electrons in a redox reaction. Those given off in the oxidation half reaction are taken up by another species in the reduction half reaction.

It is observed that in this, two species which exchanges electrons in redox reaction contains special names. The ion or molecule which accepts electrons is known as oxidizing agent as they accepts electron which results in oxidation of other species. Also, species which donates electrons are known as reducing agent as during the reaction, they reduces other species. Hence it is noted that, what is oxidized is a reducing agent while what is reduced is an oxidizing agent. An example of redox reaction is thermite reaction where iron atoms in ferric oxide loses O atoms to Al atoms thereby producing $Al_2O_3$.

$$Fe_2O_3(s)+2Al(s) \longrightarrow Al_2O_3(s)+2Fe(l)$$

The examples of oxidation-reduction or redox reactions can be traced out everytime when analyzing the reactions which is used as sources of heat or work. When natural gas burns, an oxidation-reduction reaction occurs which releases more than 800 kJ/mol of energy.

$$CH_4(g) + 2\,O_2(g) \longrightarrow CO_2(g) + 2\,H_2O(g)$$

Inside our bodies, a sequence of oxidation-reduction reaction are used to burn sugars like glucose ($C_6H_{12}O_6$) and fatty acids in fats whose reactions are shown:

$$C_6H_{12}O_6(aq) + 6\ O_2(g) \longrightarrow 6\ CO_2(g) + 6\ H_2O(l)$$

$$CH_3(CH_2)_{16}CO_2H(aq) + 26\ O_2(g) \longrightarrow 18\ CO_2(g) + 18\ H_2O(l)$$

Silver metal, is oxidized when it comes in contact with trace quantities of $H_2S$ or $SO_2$ in the atmosphere or foods, such as eggs, that are rich in sulfur compounds.

$$4\ Ag(s) + 2\ H_2S(g) + O_2(g) \longrightarrow 2\ Ag_2S(s) + 2\ H_2O(g)$$

Fortunately, the film of $Ag_2S$ that collects on the metal surface forms a protective coating that slows down further oxidation of the silver metal. Tarnishing of silver is an example of oxidation-reduction reactions which falls under corrosion. Also, similar example in series of reactions occurs when iron or steel rusts, where on heating, iron reacts with oxygen to form mixture of iron(II) and iron(III) oxides.

$$2\ Fe(s) + O_2(g) \longrightarrow 2\ FeO(s)$$

$$2\ Fe(s) + 3\ O_2(g) \longrightarrow 2\ Fe_2O_3(s)$$

Molten iron even reacts with water to form an aqueous solution of $Fe^{2+}$ ions and $H_2$ gas.

$$Fe(l) + 2\ H_2O(l) \longrightarrow Fe^{2+}(aq) + 2\ OH^-(aq) + H_2(g)$$

At room temperature, however, all three of these reactions are so slow they can be ignored. Iron corrodes at room temperature in presence of oxygen and water where in this reaction, iron is oxidized to give hydrated form of iron(II) oxide.

$$2\ Fe(s) + O_2(aq) + 2\ H_2O(l) \longrightarrow 2\ FeO\ H_2O(s)$$

As this compound has similar empirical formula as $Fe(OH)_2$, it is called iron(II), or ferrous, hydroxide. The $FeO\ H_2O$ formed in reaction is oxidized by $O_2$ dissolved in water to give a hydrated form of iron(III), or ferric, oxide.

$$4\ FeO\ H_2O(s) + O_2(aq) + 2\ H_2O(l) \longrightarrow 2\ Fe_2O_3\ 3\ H_2O(s)$$

Further it is noted that $FeO\ H_2O$ formed at metal surface will combine with $Fe_2O_3$ 3 $H_2O$ to give a hydrated form of magnetic iron oxide ($Fe_3O_4$).

$$FeO\ H_2O(s) + Fe_2O_3 \longrightarrow 3\ H_2O(s)\ Fe_3O_4\ n\ H_2O(s)$$

Since these reactions occurs in presence of water and oxygen, iron tend to rust where water collects, hence the easy way of preventing iron from rusting is to coat the metal so that it doesn't come in contact with water.

## Recognizing Oxidation-Reduction Reactions

Chemical reactions are divided into oxidation-reduction or metathesis reactions. Metathesis reactions includes acid-base reactions which involves transfer of $H^+$ ion from Bronsted acid to a Bronsted base.

| $CH_3CO_2H(aq)$ | + | $OH^-(aq)$ | $\rightleftharpoons$ | $CH_3CO_2^-(aq)$ | + | $H_2O(l)$ |
|---|---|---|---|---|---|---|
| Bronsted acid | | Bronsted base | | Bronsted base | | Bronsted acid |

It involves sharing of pair of electrons by electron-pair donor (Lewis base) and electron-pair acceptor (Lewis acid).

| $Co^{3+}(aq)$ | + | $6\ NO_2^-(aq)$ | $\rightleftharpoons$ | $Co(NO_2)_6^{3-}(aq)$ |
|---|---|---|---|---|
| Lewis acid | | Lewis base | | |

Oxidation-reduction reactions or redox reactions involves transferring of one or more electrons.

$Cu(s) + 2\ Ag^+(aq) \rightleftharpoons Cu^{2+}(aq) + 2\ Ag(s)$

They can also occur by the transfer of oxygen, hydrogen, or halogen atoms.

$$CO_2(g) + H_2(g) \rightleftharpoons CO(g) + H_2O(g)$$
$$SF_4(g) + F_2(g) \longrightarrow SF_6(g)$$

Also, there is a foolproof method of distinguishing metathesis and redox reactions which shows the reactions where no atoms undergoes a change in oxidation number are described as metathesis reactions. There is no change in the oxidation number of any atom in either of the metathesis reactions, for example.

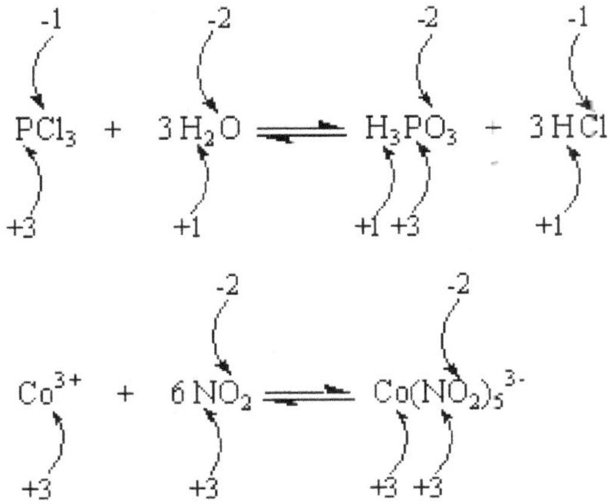


**Figure 8.2:** Metathesis Reactions.

Metathesis involves interchanging or transposition used to describe changes which occur in order of letters or sounds in a word as language develops. It occurs when word *brid* became *bird*. In chemistry, metathesis is used to describe reactions that interchange atoms or groups of atoms between molecules. When at least one atom undergoes a change in oxidation state, the reaction is an oxidation-reduction reaction. Each of the reactions in the figure below is therefore an example of an oxidation-reduction reaction.

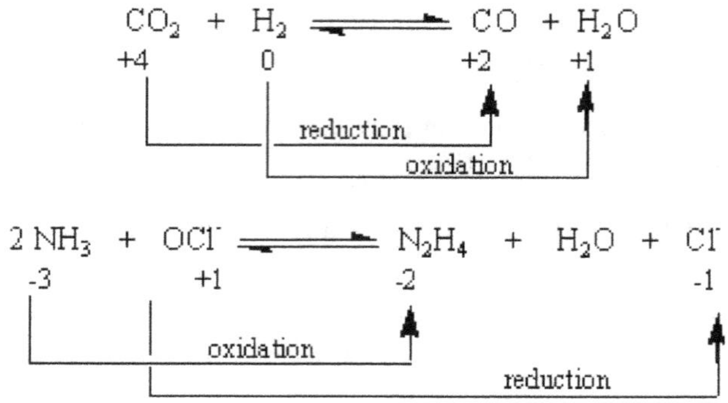


**Figure 8.3:** Oxidation-Reduction Reaction.

It is observed that:

- Oxidation is a process where one or more electrons are lost or valency of element increases.
- Reduction is a process where one or more electrons are gained or valency of element decreases.
- Oxidizing agent is a material that gain one or more electrons, i.e., valency decreases.
- Reducing agent is a material that loses one or more electrons, i.e., valency increases.
- Redox reaction involves two half reactions, one involving loss of electron or electrons (oxidation) and the other involving gain of electrons or electrons (reduction).

## 8.3 Determining Oxidation States

In order to determine the oxidation states of atoms and ions, there are certain rules which is required. First rule explains separation of ionic compound into component ions before proceeding. Second rule is total rule which should be applied as it involves charging on entire molecule or ion of concern. Third and Fourth rules are applied in that order till only one element is left undefined, hence rule 2 is applied to determine remaining oxidation number.

Unless an element has oxidation number as zero, its oxidation number is displayed with both its sign and magnitude. There are certain legitimate forms for oxidation numbers as 0, +1, +2, –1, –2, while there are no legitimate forms for oxidation numbers as –, +, 1, 2, etc.

## Assigning known Charges on Atoms or Polyatomic Ions

Always separate ionic species in their component ions before continuing. As Na in salts which always has a charge of +1, so oxidation number is +1. If anion is chloride, it will have –1 charge, and hence have an oxidation number –1.

Further, $Pb(SO_4)_2$ has 2 sulfate anions $SO_4^{2-}$, so cations must be $Pb^{4+}$ as per the rule where sum of all charged ions in ionic compound should sum to 0.

Overall Charge = sum of all oxidation numbers

It shows that sum of all oxidation numbers on all atoms in a compound or ion must equal the total charge on that compound or ion:

$H_2O$: sum of all oxidation numbers = 0 [i.e., 2 × (ox. no. H) + 1 × (ox. no. O) = 0]

$SO_4^{2-}$ : sum of all oxidation numbers = –2 [i.e., 1 × (ox. no. S) + 4 × (ox. no. O) = –2]

The result of below reactions result directly by applying rule 2 to substances containing 1

Element:

a) The oxidation number on each atom in a pure element is 0.

$H_2$: oxidation number on each H = 0
$S_8$: oxidation number on each S = 0
Mg: oxidation number on each Mg = 0

b) The oxidation number on each atom in a monatomic ion is equal to its charge.

$Mg^{2+}$: oxidation number on Mg = +2

$Cl^-$ : oxidation number on Cl = –1

c) Polyatomic species containing a single element will share the total charge equally.

$Hg_2^{2+}$ : oxidation number on each Hg = +1 [I.e., 2 × (ox. no. Hg) = +2]

$O_2^{-1}$ : oxidation number on each O = –½ [I.e., 2 × (ox. no. O) = –1] [This is the superoxide anion]

## Oxidation Number on H (in compounds)

It is noted that normally the oxidation number on H compound is +1, till it is bound to a metal to make a hydride, $MH_n$, where its charge and makes the oxidation number as –1. Here the only other exception is in pure element $H_2$, where it is forced to be 0 as per first rule.

HCl: oxidation number on H = +1 [Thus, ox. no. on Cl = –1]
$CH_4$: oxidation number on H = +1 [Thus, ox. no. on C = –4]
$MgH_2$: oxidation number on H = –1 [Thus, ox. no. on Mg = +2]

### Oxidation number on O (in compounds)

It is noted that normally oxidation number on O compound is –2 as the only exceptions other than pure elemental forms of $O_2$ or $O_3$, where it is 0 are those where it is forced to be different as:

a) The ox. no. of O in peroxides, (compounds with $O_2^{-2}$ ion), is –1: e.g. $H_2O_2$, $Na_2O_2$, $MgO_2$, etc.

b) The ox. no. of O in superoxides, (compounds with $O_2^-$ ion), is –½: e.g. $KO_2$, $RbO_2$, etc.

c) The ox. no. of O in $OF_2$ is +2; F has to be -1.

Here, all exceptions given above for oxygen, O, be obtained using second rule, after assigning oxidation number of other elements in formula as in case of $MgO_2$, Mg is in group 2 having +2 charge, hence second rule shows the following:

Total charge = 0 = (ox. no. Mg) + 2(ox. no. O) = +2 + 2(ox. no. O).

On substituting oxidation number on Mg will give +2 and solving for oxygen gives (ox. no. O) = –1. Hence oxygen's oxidation number is already been assigned after applying rules 1 and 2, so exceptions mentioned in fourth rule (a) and (b) are not required. Whenever an oxidation number on oxygen cannot be determined from other elements, then it is always assigned a value of –2.

Fourth rule (c) arises from fact that fluorine is more electronegative than oxygen, hence assign F to have oxidation number equal to normal charge as –1 since it is in group 17. Applying second rule to give oxidation number a +2 to O.

Use these rules in required manner as:

- Apply First rule if substance is ionic and separate it into its component ions and treat each ion separately.
- Apply Second – Fourth rules on each ion which are independent of other and set the equation in second rule.
- Apply Third and Fourth rules in the order till only single element is left unassigned and apply second rule equation to solve for sole remaining oxidation number.

## 8.3.1 Solving Atoms and Ions

### $FeSO_4$

It is an ionic substance having sulfate anion, $SO_4^{2-}$, As charges on the ions will sum to 0, the charge on Fe, should be +2. To find the values for remaining elements, set the second rule equation:

$$-2 = 1 \times (\text{ox. no. S}) + 4 \times (\text{ox. no. O}).$$

This equation has two unknowns, so applying the rules. As there are no H atoms, so using rule fourth and set oxidation number on O as –2. This will give 1 for S and use the result for oxygen, as

$$-2 = (\text{ox. no. S}) + 4(-2)$$

On solving, we have ox. no. S = +6.

**$CH_4O$**

It is not an ionic substance hence leaving rule first. Now rule second will give the following equation:

$$0 = 1 \times (\text{ox. no. C}) + 4 \times (\text{ox. no. H}) + 1 \times (\text{ox. no. O})$$

On applying third rule and setting the ox. no. on H to +1 as two unknowns are there, so applying rule fourth and set ox. no. on O to –2 which leaves 1 unknown for C, so

$$0 = (\text{ox. no. C}) + 4 \times (+1) + 1 \times (-2)$$

On solving, we get (ox. no. C) = –2.

**$N_2O_4$**

It is not an ionic substance hence leaving rule first. Now rule second will give the following equation:

$$0 = 2 \times (\text{ox. no. N}) + 4 \times (\text{ox. no. O})$$

Since there are no H atoms, so skipping fourth rule and set ox. number on O to –2, so

$$0 = 2 \times (\text{ox. no. N}) + 4 \times (-2) = 2(\text{ox. no. N}) - 8.$$

On solving, (ox. no. N) = + 4.

**$Cr_2O_7^{-2}$**

In this, rule first is not required as the ions have already been separated, so applying second rule 2:

$$-2 = 2 \times (\text{ox. no. Cr}) + 7 \times (\text{ox. no. O})$$

The ox. no. on O is set to –2 as per fourth rule, hence on substituting:

$$-2 = 2 \times (\text{ox. no. Cr}) + 7 \times (-2) = 2\,(\text{ox. no. Cr}) - 14$$

On solving, (ox. no. Cr) = +6.

**$Fe(CH_3COO)_3$**

In this, application of first rule shows the presence of acetate anion $CH_3COO^-$. For balancing the charge, charge on Fe should be +3. To find oxidation numbers on remaining elements, use second rule to acetate anion:

$-1 = 2 \times (\text{ox. no. C}) + 3 \times (\text{ox. no. H}) + 2 \times (\text{ox. no. O})$

Since there are three unknowns, so using third and fourth rules, so oxidations of +1 on H and –2

on O, hence second rule equation will become:

$-1 = 2 \times (\text{ox. no. C}) + 3 \times (+1) + 2 \times (-2) = 2 \times (\text{ox. no. C}) + 3 - 4 = 2 \times (\text{ox. no. C}) - 1$

This gives

$0 = 2\,(\text{ox. no. C})$

Hence, (ox. no. C) = 0.

**$NH_4NO_2$**

Using the first rule as this compound has ammonium ion, $NH_4^+$ and nitrite anion, $NO_2^-$ so

applying remaining rules to each ion separately:

**$NH_4^+$**

Using the second rule, so equation gives

$+1 = 1 \times (\text{ox. no. N}) + 4 \times (\text{ox. no. H})$

Using third rule 3, so equation gives (ox. no. H) = +1 and further becomes

$+1 = 1 \times (\text{ox. no. N}) + 4 \times (+1) = (\text{ox. no. N}) + 4$

On solving, (ox. no. N) = –3

**$NO_2^-$**

Applying second rule, so equation will lead to:

$-1 = 1 \times (\text{ox. no. N}) + 2 \times (\text{ox. no. O})$

Using fourth rule gives (ox. no. O) = –2, and equation becomes

$-1 = 1 \times (\text{ox. no. N}) + 2 \times (-2) = (\text{ox. no. N}) - 4$

On solving, (ox. no. N) = +3

Here, oxidation number on N is –3 in $NH_4^+$ and is +3 in $NO_2^-$. $NH_4NO_3$ is a substance where different oxidation numbers occur on atoms from same element. It occurs when atoms occur in different ions where oxidation number on element in cation is different from oxidation number on same element in anion. Such phenomenon will never take place among atoms from identical elements in similar cation or in similar anion or in non-ionic compound.

## 8.4 The pE Concept

pE is the negative logarithm of electron concentration (-log[e]) in a solution that is directly proportional to redox potential. It is used as a unit of reduction potential instead of Eh in environmental chemistry. On normalizing pE of hydrogen to 0, a relation pE=16.9 Eh at room temperature exists which is good for understanding redox potential where transfer of electrons instead of absolute concentration of free electrons in thermal equilibrium shows show one usually thinks of redox potential.

In environmental chemistry, reduction potential determines when oxidizing or reducing conditions are common in water or soil which helps in predicting states of different chemical species in water. pE values in water range from -12 to 25; the levels where the water itself becomes reduced or oxidized.

The reduction potentials in natural systems lie near the boundaries of stability region of water. Aerated surface water, rivers, lakes, oceans, rainwater and acid mine water have oxidizing conditions. Places with limitations of air like submerged soils, swamps and marine sediments, reducing conditions are norm. Intermediate values are rare and usually a temporary condition found in systems moving to higher or lower pE values.

Normally in environmental situations, it is easy to have complex non-equilibrium conditions exists among large number of species which is not possible for accurate and precise measurements of reduction potential. It is possible to obtain correct value and conditions as being in oxidizing or reducing regime.

It is not highly correct as the availability of electrons in terms of electron activity where electrons are treated like $H^+$ having artificial activity where electron activity is written as **pe**

$$- \log (e-) = pe$$

With this notation, an electron is treated as any other reactant. Now consider a generalized half reaction to show the convention and utility as:

$$OX + n\ e- = RED$$

$$K = (RED) / (OX)\ (e-)^n$$

$$\log K = \log (RED)/OX - n \log e-$$

$$= \log (RED)/OX + npe$$

$$\log K - npe = \log (RED)/(OX)$$

In this, we can measure pe and know the ratio of RED/OX. Here it is seen that "p" unit is $\log_{10}$-based quantity, so we can report the activity of any chemical species in p units whose formula is:

$$pX = -\log_{10} A_x$$

Hence, pH is

$$pH = -\log_{10} A_{H+}$$

which for low concentration solutions will be:

$$pH \approx -\log_{10}[H^+]$$

The p*E* is -$\log_{10}$ of electron activity, so

$$pE = -\log_{10} A_{e-}$$

Here pE is related to reduction potential by *FRT* factor of Nernst equation, so pE will be:

$$pE = \frac{E}{0.0591} @ 25°C$$

The Nernst equation is normally applied to show p*E* which can be recast using p*E* formalism

$$pE = pE - \frac{1}{n}\log_{10} Q$$

that defines p$E^0$

**Important**

Valence: Atoms have different valences (degree of negative or positive charge) determined by whether they have enough protons to balance their electrons. Elements and molecules have tendencies to associate or dissociate with electrons (change valence)

**Reduction:** Gaining electrons (valence becomes more negative). A reductant is a compound that donates electrons

**Oxidation:** Losing electrons (valence becomes more positive). An oxidant is a compound that accepts electrons "LEO the lion says GER"

**Redox potential:** It is an arbitrary scale showing how to oxidized or reduced things which are compared to a standard

$$H^+ + e^- \leftrightarrow \frac{1}{2} H_{2(g)}$$

Electron Flow: Electron flow relative to hydrogen standard. As Eh decreases, the solution is more reduced (has more electrons to give) and as Eh increases, the solution is more oxidized (will accept electrons).

**Electron Activity: pE Electron activity of solution**

- pE=-log[e-]

- measured at pH=7; =16.9 Eh (so is related to electron potential) Large and positive in strongly oxidizing solutions (low electron activity)

Redox reactions control the form of elements and the distribution of different forms of elements in lakes

**Iron**

A. Forms of oxygen

1. $Fe^{2+}$ (ferrous) < ~300 mv Eh
2. $Fe^{3+}$ (ferric) > ~300 mv Eh

B. Iron in oxygenated waters

C. Iron in anoxic waters

D. Significance:

1. $Fe^{3+}$ binds with P, element abundances are interrelated
2. as a limiting nutrient

Redox 'Battlefield': Many reactions are mediated by bacteria

A. Players in the 'battle'

- One side – PS organisms take light energy and make reduced compounds (local chemical disequilibrium)
- Other side – using these reduced compounds – heterotrophs extracting potential energy and attempting to reestablish equilibrium

**Reactions**

The free energy depends on the energy difference between the reactants and the products.

**Gibbs free energy** of reaction - Determines which reactions are most energetically favorable

$$\Delta G^\circ = -RT \ln K_{eq}$$

$\Delta G^\circ$ = standard free energy of formation
R=gas constant (8.314 J $K^{-1}$ $mol^{-1}$)
$K_{eq}$=equilibrium constant
T=temperature (K)

$$= -nF\Delta E$$

n=number of electrons involved in the reaction
F=Faraday=23 kcal/mol
$\Delta E$=difference in redox potential in reaction

$\Delta G^{\circ\prime}$ = Sum (delta$G^{\circ\prime}$products) - Sum (delta$G^{\circ\prime}$reactants )

The minimum energy necessary to be useful to organisms is deltaG°=-7kcal/mol – that is what is necessary or ATP formation

**Redox Reactions**

**1. Aerobic Respiration (oxidation of organic matter)**

a) reductant CHO, oxidant $O_2$

b) Reaction:

$$C_6H_{12}O_6 + 6O_2 \leftrightarrow 6CO_2 + 6H_2O$$

c) delta G° = -686 kcal/mol

**2. Nitrogen**

a) Nitrate Reduction

(i) Reductant CHO, oxidant $NO_3^-$

(ii) Reaction: Dissimilatory Nitrate Reduction

$$CH_2O + NO_3^- + 2H^+ \leftrightarrow CO_2 + \frac{1}{2}N_2 + 2H_2O$$

(iii) $\Delta G^{o\prime}$ = -649 kcal/mol

(iv) This is denitrification, reducing nitrate, $NO_3^-$, to $N_2$ gas

b) Assimilatory nitrate reduction

(i) Uptake of nitrate by an organism

(ii) Reaction: $NO_3^- + H^+ \rightarrow NH_3 - organic$

c) Nitrification – production of nitrate

(i) Reactions:

a) $NH_4^+ + \frac{3}{2}O_2 \leftrightarrow NO_2^- + 2H^+ + H_2O$ $\Delta G°$ = -65.7 kcal/mol

b) $NO_2^- + \frac{1}{2}O_2 \leftrightarrow NO_3^-$ $\Delta G°$ = -17.5 kcal/mol

(ii) So $\Delta G°1$ = -65.7 kcal/mol and $\Delta G°2$ = -17.5 kcal/mol

(iii) This produces the source material for denitrification

**3. Iron**

**Iron Oxidation**

a) Reductant $Fe^{2+}$, oxidant $O_2$

b) Reaction:

$$Fe^{2+} + \frac{1}{4}O_2 + H^+ \leftrightarrow Fe^{3+} + \frac{1}{2}H_2O$$

$$Fe^{3+} + 3H_2O \leftrightarrow Fe(OH)_3 + H^+$$

c) ΔG°′ = -10.6 kcal/mol

**Iron Reduction**

a) Reaction: $Fe(OH)_3 + 3H^+ \rightarrow Fe^{2+} + 3H_2O$

b) Can occur with some oxygen present, but not energetically favorable

c) ΔG°′ = -300 kcal/mol

**4. Sulfur**

**Sulfate Reduction**

a) Reductant CHO, oxidant $SO_4^{2-}$

b) Reaction: $2CHO + SO_4^{2-} + 3H^+ \leftrightarrow 2CO_2 + HS^- + 2H_2O$

c) ΔG°′ = -190 kcal/mol

**Sulfide Oxidation**

a) reductant $HS^-$, oxidant $O_2$

b) Reaction: $HS^- + 2O_2 \leftrightarrow SO_4^{2-} + H^+$

c) ΔG°′ = -190 kcal/mol

**5. Methane**

**Methanogenesis**

a) Reductant $H_2$, oxidant $CO_2$

b) Reaction: $4H_2 + CO_2 \leftrightarrow CH_4 + 2H_2O$

c) ΔG°′ = -8.3 kcal/mol

**Methane Oxidation**

a) Reductant $CH_4$, oxidant $O_2$

b) Reaction: $CH_4 + 2O_2 \leftrightarrow CO_2 + 2H_2O$

c) ΔG°′ = -193.5 kcal/mol

d) Whether methane can be oxidized with either $SO_4^{2-}$ or $NO_3^-$ is disputed.

# 9 Carbon Dioxide

## 9.1 Significance and Importance

Carbon dioxide is a molecule with the molecular formula $CO_2$. The linear molecule consists of a carbon atom that is doubly bonded to two oxygen atoms, O=C=O. Although carbon dioxide mainly consists in the gaseous form, it also has a solid and a liquid form. It can only be solid when temperatures are below -78°C. Liquid carbon dioxide mainly exists when carbon dioxide is dissolved in water. Carbon dioxide is only water-soluble, when pressure is maintained. After pressure drops the $CO_2$ gas will try to escape to air. This event is characterised by the $CO_2$ bubbles forming into water.

Carbon dioxide is one of the most abundant gasses in the atmosphere. Carbon dioxide plays an important part in vital plant and animal process, such as photosynthesis and respiration. Green plants convert carbon dioxide and water into food compounds, such as glucose, and oxygen. This process is called photosynthesis. The reaction of photosynthesis is as follows:

$$6\ CO_2 + 6\ H_2O \rightarrow C_6H_{12}O_6 + 6\ O_2$$

Plants and animals, in turn, convert the food compounds by combining it with oxygen to release energy for growth and other life activities. This is the respiration process, the reverse of photosynthesis. The respiration reaction is as follows:

$$C_6H_{12}O_6 + 6\ O_2 \rightarrow 6\ CO_2 + 6\ H_2O$$

Carbon dioxide dissolves slightly in water to form a weak acid called carbonic acid, $H_2CO_3$, according to the following reaction:

$$CO_2 + H_2O \rightarrow H_2CO_3$$

After that, carbonic acid reacts slightly and reversibly in water to form a hydronium cation, $H_3O^+$, and the bicarbonate ion, $HCO_3^-$, according to the following reaction:

$$H_2CO_3 + H_2O \rightarrow HCO_3^- + H_3O^+$$

This chemical behaviour explains why water, which normally has a neutral pH of 7 has an acidic pH of approximately 5.5 when it has been exposed to air.

Carbon dioxide does dissolve in water, however the system is somewhat complex. First the $CO_2$ dissolves according to:

$$CO_2\ (g) \Leftrightarrow CO_2\ (l)$$

At room temperature, the solubility of carbon dioxide is about 90 $cm^3$ of $CO_2$ per 100 ml water ($c_l/c_g$ = 0.8). Any water-soluble gas becomes more soluble as the temperature decreases, due to the thermodynamics of the reaction: GAS (l) ⇔ GAS (g). The entropy change, ΔS, of this reaction is positive because the gas molecules are less constrained than the gas molecules in solution. The change in Free energy of reaction with an increase in temperature is -ΔS. This effect is particularly large for gases like $CO_2$ that undergo specific reactions with water.

Equilibrium is established between the dissolved $CO_2$ and $H_2CO_3$, carbonic acid.

$$CO_2\ (l) + H_2O\ (l) \Leftrightarrow H_2CO_3\ (l)$$

This reaction is kinetically slow. At equilibrium, only a small fraction (ca. 0.2 - 1%) of the dissolved $CO_2$ is actually converted to $H_2CO_3$. Most of the $CO_2$ remains as solvated molecular $CO_2$. As equation:

$$K_r = \frac{[H_2CO_3]}{[CO_2]} \approx 1.7.10^{-3}$$

In fact, the $pK_a$ most reported for carbonic acid ($pK_{a1}$ = 6.37) is not really the true $pK_a$ of carbonic acid. Rather, it is the $pK_a$ of the equilibrium mixture of $CO_2$ (l) and carbonic acid. Carbonic acid is actually a much stronger acid than this, with a true $pK_{a1}$ value of 3.58. However these values are also temperature dependent. Carbonic acid is a weak acid that dissociates in two steps.

$$H_2CO_3 + H_2O \Leftrightarrow H_3O^+ + HCO_3^-\ pK_{a1}\ (25\ °C) = 6.37$$

$$HCO_3^- + H_2O \Leftrightarrow H_3O^+ + CO_3^{2-}\ pK_{a2}\ (25\ °C) = 10.25$$

Note that these carbonate anions can interact with the cations present in the water to form insoluble carbonates. For instance, if $Ca^{2+}$ is present limestone, $CaCO_3$ is formed and if $Mg^{2+}$ is present $MgCO_3$ is formed. The formation of these deposits is an additional driving force that can pull the equilibrium more to the right resulting in acidification of the water.

$$Ca^{2+} + CO_3^{2-} \Leftrightarrow CaCO_3\ S = 4.96 \times 10^{-9}\ (S = \text{solubility constant})$$

$$Mg^{2+} + CO_3^{2-} \Leftrightarrow MgCO_3\ S = 6.82 \times 10^{-6}$$

The above presented more schematically:

$$CO_2(g) \Leftrightarrow CO_2\ (l) \overset{+H_2O}{\Leftrightarrow} H_2CO_3 \overset{+H_2O}{\Leftrightarrow} \underset{+H_3O^+}{HCO_3^-} \overset{+H_2O}{\Leftrightarrow} \underset{+H_3O^+}{CO_3^{2-}} \overset{+Ca^{2+}}{\Leftrightarrow} CaCO_3$$

Note that the reverse is also true and that the scheme represents the solubility of $CaCO_3$ in an acidic solution resulting in the liberation of $CO_2$ in the atmosphere.

**Deriving [$H_2CO_3$]**

If we assume $CO_2$ is a simple gas we can apply Henry's law that describes the equilibrium between vapor and liquid. Thus:

$$p_{CO_2} = K \cdot x_{CO_2}$$

where

$p_{CO_2}$ = partial pressure of gas in bulk atmosphere (Pa)

K = constant (Pa)

$x_{CO2}$ = equilibrium mole fraction of solute in liquid phase

The solubility of $CO_2$ is temperature dependent as shown in Table 9.1.

**Table 9.1:** Solubility of $CO_2$ at a Partial Pressure for $CO_2$ of 1 bar abs

| Temperature (°C) | 0 | 10 | 20 | 30 | 40 | 50 | 80 | 100 |
|---|---|---|---|---|---|---|---|---|
| Solubility (cm³ $CO_2$/g water) | 1.8 | 1.3 | 0.88 | 0.65 | 0.52 | 0.43 | 0.29 | 0.26 |

Furthermore, as stated above, $CO_2$ reacts with the water on dissolution and therefore one would expect that Henry's law has to be modified. However, a form of Henry's law can be used for modeling the solubility of carbon dioxide in water for pressures up to about 100 MPa, as can be seen in figure.

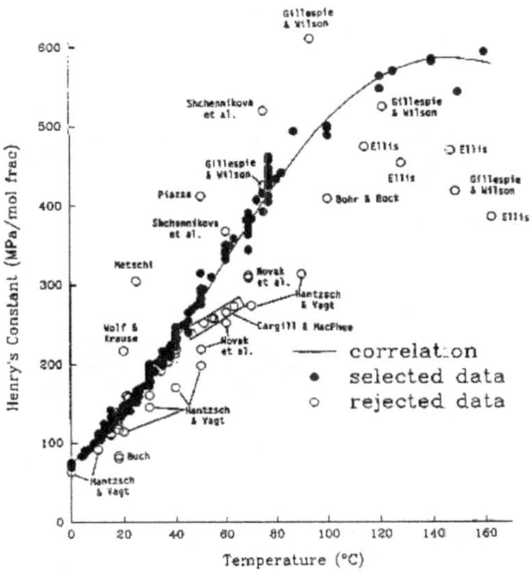


**Figure 9.1:** Henry's Constant for Carbon Dioxide in Water.

As per Krichevsky-Kasarnovsky Equation derived from Henry's Law, the equation is used to model system $CO_2$-$H_2O$ at temperatures below 100 °C. Hence in range of interest, 20-35 °C, the Henry coefficient for $CO_2$ in water goes from 150 - 200 MPa/mole fraction. Applying the above to the conditions under investigation:

Temperature range: 20 – 35 °C

Pressure range: 80 – 90 bar

$CO_2$ concentration in gas phase: 1.3-1.7 mol%

The partial pressure of $CO_2$ in the gas phase is therefore in the range:

$$1.3/100 * 80 * 0.1 = 0.104 \text{ MPa}$$

$$1.7/100 * 90 * 0.1 = 0.153 \text{ MPa}$$

Applying Henry's Law we calculate a $CO_2$ mole fraction in water in the range:

$$x_{low} = 0.104 / 200 = 0.00052$$

$$x_{high} = 0.153 / 150 = 0.00102$$

Converting mole fractions to concentrations:

At 20 °C the molar density of water = 998.21/18.02 = 55.39 mol/l

At 35 °C the molar density of water = 994.37/18.02 = 55.18 mol/l

Thus the $CO_2$ concentration range in water under these conditions is:

$$c_{low} = 0.00052 * 55.18 = 0.029 \text{ mol/l}$$

$$c_{high} = 0.00102 * 55.39 = 0.056 \text{ mol/l}$$

Calculating the pH

The basic equations needed to calculate the pH are derived from equation (3) and (4).

Note: $x = [H_3O^+]$ and $y = [OH^-]$

The protolysis constants:

$$K_{a1} = \frac{x[HCO_3^-]}{[H_2CO_3]} \approx \left( \frac{x[HCO_3^-]}{[CO_2 + H_2CO_3]} \approx \frac{x[HCO_3^-]}{[CO_2]_l} \right)$$

$$K_{a2} = \frac{x[CO_3^{2-}]}{[HCO_3^-]}$$

Assuming the initial concentration of carbonic acid = c then we derive the steuchiometric relation:

$$c = [H_2CO_3] + [HCO_3^-] + [CO_3^{2-}]$$

The water constant: x.y = $K_w$ and $pK_w$ =14

The electro neutrality equation:

$$2\,[CO_3^{2-}] + [HCO_3^-] + y = x$$

We now have 5 equations and 5 unknowns. From the equilibrium constants we derive

$$[HCO_3^-] = \frac{x[CO_3^{2-}]}{K_{a2}}$$

and

$$[HCO_3^-] = \frac{x[CO_3^{2-}]}{K_{a2}}$$

Substituting these into the steuchiometric relation we derive:

$$c = \frac{x^2[CO_3^{2-}]}{K_{a1}K_{a2}} + \frac{x[CO_3^{2-}]}{K_{a2}} + [CO_3^{2-}] = [CO_3^{2-}]\left(1 + \frac{x}{K_{a2}} + \frac{x^2}{K_{a1}K_{a2}}\right) \Rightarrow$$

$$[CO_3^{2-}] = \frac{K_{a1}K_{a2}.c}{x^2 + K_{a1}.x + K_{a1}K_{a2}}$$

$$[HCO_3^-] = \frac{K_{a1}.x.c}{x^2 + K_{a1}.x + K_{a1}K_{a2}}$$

These equations can be used to calculate in which pH area which $CO_2$ species dominates as can be seen in Figure 9.2.

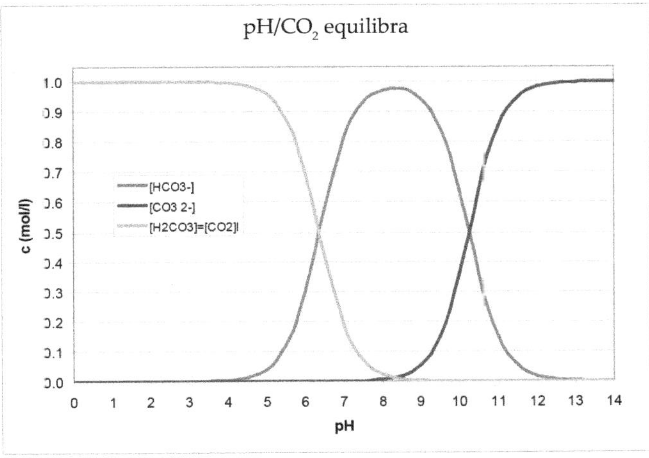


**Figure 9.2:** pH and $CO_2$ Species.

Substituting these into the electro neutrality equation:

$$2\frac{K_{a1}K_{a2}.c}{x^2+K_{a1}.x+K_{a1}K_{a2}}+\frac{K_{a1}.x.c}{x^2+K_{a1}.x+K_{a1}K_{a2}}+y=x\Rightarrow$$

$$c.\frac{2.K_{a1}K_{a2}+K_{a1}.x}{x^2+K_{a1}.x+K_{a1}K_{a2}}+y=x$$

Combining the above with Kw=x.y we derive a 4[th] degree equation in x:

$$c.\frac{2.K_{a1}K_{a2}+K_{a1}.x}{x^2+K_{a1}.x+K_{a1}K_{a2}}+\frac{K_w}{x}=x$$

Compare:

$$\frac{2.K_{a1}K_{a2}+K_{a1}.x}{x^2+K_{a1}.x+K_{a1}K_{a2}}$$

with

$$\frac{K_{a1}}{K_{a1}+x}+\frac{K_{a2}}{K_{a2}+x}\Rightarrow\frac{2.K_{a1}K_{a2}+K_{a1}.x+K_{a2}.x}{x^2+K_{a1}.x+K_{a2}.x+K_{a1}K_{a2}}$$

If $K_{a1} >> K_{a2}$ then $K_{a1}.x >> K_{a2}.x$ and we can consider the equations equal and derive for the electro neutrality equation:

$$\frac{K_{a1}}{K_{a1}+x}c+\frac{K_{a2}}{K_{a2}+x}c+\frac{K_w}{x}=x$$

Estimating the difference between Ka and x can help simplifying the equation even further.

Furthermore: pH = -log x. Applying the equations derived above to the $CO_2$ concentrations calculated we calculate by applying that if Ka << x a valid approach if the pH is ca. 4 we can write:

$$\frac{K_{a1}}{x}c+\frac{K_{a2}}{x}c+\frac{K_w}{x}=x\Rightarrow x=\sqrt{K_{a1}.c+K_{a2}c+K_w}$$

Thus with the $CO_2$ concentration range in water calculated above we calculate for the pH range:

$c_{low}$ = 0.029 mol/l ➔ pH = 3.95

$c_{high}$ = 0.056 mol/l ➔ pH = 3.81

Verification has been done using the basic equation via a numeric solution in excel employing the goal seek function. The same pH values as derived above are found. It is possible to refine the result even further. The dissociation constant is also depending on the temperature of the solution.

**Table 9.2**: Dissociation Constant ($K_{1A}$) of Carbonic Acid at Various Temperatures

| Temperature (°C) | 0 | 5 | 10 | 15 | 20 | 25 | 30 | 35 | 40 | 45 | 50 |
|---|---|---|---|---|---|---|---|---|---|---|---|
| $K_{1A}.10^7$ | 2.64 | 3.04 | 3.44 | 3.81 | 4.16 | 4.45 | 4.71 | 4.90 | 5.04 | 5.13 | 5.19 |

## 9.2 Dependent Carbonate Equilibrium

### 9.2.1 Carbonate

Calcite ($CaCO_3$) is one of the most common minerals at or near the surface of Earth and thus is one of the main contributors to carbonate geochemistry. Less common carbonate minerals include dolomite ($MgCa(CO_3)_2$), aragonite, a calcite polymorph, azurite and malachite (copper hydroxycarbonate minerals), siderite ($FeCO_3$) and rhodochrosite ($MnCO_3$) which can be important spatially and economically.

Although calcite does occur in igneous and metamorphic rocks, the mineral is present dominantly in the sedimentary environment which is the focus of this article. Calcite may be precipitated to form a relatively pure mono-mineral rock, be present as a cement binding sediments into indurate rock, occurs simply as a trace mineral in rocks and composes many shells and fossils. Weathering of calcite occurs relatively easy and depends largely on the amount of carbonic acid present. Other carbonate minerals, with the exception of dolomite, generally are dissolved and precipitated in a manner geochemically similar to calcite.

#### Calcite Solubility Products

The solubility of $CaCO_3$ depends on which polymorph is considered and the reference for the equilibrium constant. For example, at 25°C the ion concentration products ($K_{sp}$) for calcite and aragonite (the most common polymorphs of $CaCO_3$) are $10^{-8.48}$ and $10^{-8.34}$, respectively. It is given that calcite and aragonite $K_{sp}$ values of $10^{-8.35}$ and $10^{-8.22}$ respectively where the $K_{sp}$ values for calcite solubility are 35% more than those from APHA. A similar relationship exists for aragonite showing $K_{sp}$ values for aragonite that are 31% larger. Variation in solubility experiments can yield significant differences in the $K_{sp}$.

#### Temperature

In an increase of mineral solubility with increasing temperature related to minerals, many carbonate minerals are soluble in cold water. The $K_{sp}$ for calcite at 0°C and at 50°C is $10^{-8.02}$ and $10^{-8.63}$, respectively. Such values show four fold difference in calcite solubility due to temperature. A simple explanation for carbonate minerals of being soluble in cold water is that the dissolution reaction for carbonate minerals is exothermic that appears in high temperatures favoring solid phase over dissolved ions.

With increase in solubility products and decreasing temperature, carbon dioxide is more soluble at lower temperatures, further favoring carbonate mineral dissolution in cooler environments. More carbonic acid is present in cold water at a given partial pressure of carbon dioxide ($PCO_2$) and carbonic acid concentration is the controlling factor for the solubility of carbonate minerals under natural conditions. Carbon dioxide is obtained from the air or from decomposition of organic matter that releases carbon dioxide which reacts with water to form carbonic acid. These factors combine to form the calcite compensation depth (about 4,500 m) below which calcite dissolves.

When geothermal waters reach the surface of Earth and precipitate tufa, a calciumcarbonate rock, some reaction must overcome the temperature decrease or these waters would dissolve calcite rather than deposit tufa as the water cools. The required reaction is the loss of $CO_2$ due to lower $PCO_2$ pressure at Earth's surface and subsequent a decrease in the amount of carbonic acid, which controls the amount of dissolved calcite.

**Pressure**

Increased partial pressure of $CO_2$ near the surface of Earth increases the amount of $CO_2$ that dissolves in water, therefore, increasing the solubility of carbonate minerals such as calcite. Pressure alone does not affect the solubility of calcite as much as the effect of temperature. Nonetheless, where pressure is large, for example, deep within the ocean, its effect alone can increase calcite solubility about two fold.

**Ionic Strength**

The presence of other ions in solution shields $Ca^{2+}$ and $CO_3^{2-}$ ions from interacting and precipitating. As described, this situation results as activity of ions in electrolyte solutions which is less than concentration where interference from other ions increases concentrations and charges. Activity defined as concentration of ion at zero ionic strength which affects concentration and charge of ions shown by ionic strength (I) of solution as

$$I = 0.5\ \Sigma\ m_i z_i^{\ 2}$$

where

$m_i$ = molar concentration of an ion ($i$)

z = charge of ion

Ionic strength can be used to calculate the activity coefficient ($\gamma$) which relates molar concentration to activity and $\gamma$ is 1 or less.

$$m_i\ (\gamma) = a$$

where $a$ is the activity of the ion.

The $\gamma$ can be calculated from the Debye-Hückel equation or other similar equations that make use of I and some constants that relate to the dielectric constant of water and temperature (A and B) and effective size (a) of ions. Values of the constants can be obtained from Faure.

$$-\log \gamma = \frac{Az^2 I^{1/2}}{I + aBI^{1/2}}$$

**Saturation Index**

Waters tend to precipitate $CaCO_3$ when oversaturated with respect to $CaCO_3$ and tend to dissolve $CaCO_3$ if unsaturated with respect to $CaCO_3$. The Saturation Index is perhaps the most widely used method to find amount of calcite that will be precipitated or dissolved. Many computer codes calculate RS (relative saturation) which is related to SI as shown below:

SI = log RS

where

RS = ratio of $CaCO_3$ activity product (Ka) to $CaCO_3$ solubility product (Ksp)

SI values of 1 indicate saturation, negative values indicate undersaturation and positive values indicate oversaturation.

**Measured Alkalinity**

Alkalinity is the measure of the acid consuming ability (power) of a solution and is expressed as mg/L $CaCO_3$. In nature the main ions that neutralize $H^+$ ions are $HCO_3$ and $CO^2_3$ Although $OH^-$, $PO^{3-}_4$ and other negative ions can neutralize $H^+$ ions, generally the concentrations of these ions is so small in the natural environment that $HCO_3$ and $CO^{2-}_3$ are the only significant acid neutralizing ions. In some situations, especially in anthropogenic influenced situations $OH^-$ can be important, hence $OH^-$ is included in the following calculations.

Alkalinity is usually determined by titrating a 50 or 100 mL water sample with 0.02 N $H_2SO_4$ to a pH of 8.3 for $HCO_3$ and/or $OH^-$. This 8.3 pH end point is often referred to as the phenolphthalein (or P) end point because this indicator changes colors at a pH of 8.3. Titration continues past 8.3 to a pH of about 4.5 as any $HCO_3$ is neutralized. The pH of 4.5 can vary from 4.3-4.9 as per presence and amounts of $CaCO_3$ and certain anions. This final end point is referred to as total alkalinity. These pH titration end points are based on inflection points on the titration curve of pH versus milliliters of acid added to the water sample titration curve for neutralization of the bases. The amount of acid necessary to reach the phenolphthalein end point is equivalent to the amount of base ($CaCO_3$) neutralized and is referred to as P alkalinity.

The amount of $CaCO_3$ to completely neutralize $_H^+$ ions added is referred to as T alkalinity. Thus, T must be larger than or equal to P. There are five alkalinity conditions possible:

$HCO_3$

$CO^{2-}_3$

$OH^-$

$HCO^-_3 + CO^{2-}_3$

$CO^{2-}_3 + OH^-$

Bicarbonate and $OH^-$ ions cannot co-exist under typical natural conditions because these ions react to form $CO^{2-}_3$ ions and $H_2O$. Such situations are described as:

**Bicarbonate Alkalinity**

If initial pH of a sample is less than 8.3, there is only bicarbonate alkalinity and no P alkalinity. Or one can state that there is only T alkalinity.

**Carbonate Alkalinity**

There are two abrupt changes in pH during alkalinity titration, one at 8.3 and the other at 4.5. If only $CO^{2-}_3$ ions are present, amount of acid necessary to lower the pH below 8.3 is ½ the total acid (i.e., T) necessary to lower the pH below 4.5. The first abrupt pH change at 8.3 represents the conversion of carbonate to $HCO_3$ and the second abrupt pH change at 4.5 represents the neutralization of the $HCO_3$ created from carbonate. If only $CO^{2-}_3$ ions are present T = 2P.

**Hydroxide Alkalinity**

If the initial pH is greater than 8.3 and the addition of acid rapidly lowers the pH below 4.5, there is only hydroxide alkalinity and thus T=P.

**Bicarbonate Plus Carbonate Alkalinity**

As P shows conversion of $CO^{2-}_3$ to $HCO_3$ the same amount of acid (P) is required to neutralize the $HCO_3$ ions formed from the $CO^{2-}_3$ ions, i.e., 2P represents complete neutralization of $CO^{2-}_3$ ions present. In addition, $HCO_3$ originally present is included in T, which means that T>2P. Because $CO^{2-}_3$ ions are present, there will be two abrupt changes in pH (at 8.3 and 4.5).

**Carbonate Plus Hydroxide Alkalinity**

When $CO^{2-}_3$ and $OH^-$ ions are present, the amount of acid required to lower the pH below 8.3 is that necessary to convert $CO^{2-}_3$ to $HCO_3$ (P) plus that necessary to neutralize $OH^-$ions. Neutralization of $HCO^-_3$ ions formed from $CO^{2-}_3$ requires P amount of acid. These conditions constrain the P and T relationship to T<2P.

Table 9.3 summarizes the equations necessary to determine the portion of alkalinity due to $HCO^-_3$, $CO^{2-}_3$ ions and $OH^-$. At a pH of 9 the $OH^-$ concentration is only 1.0 mg/L as $CaCO_3$, and can generally be neglected in alkalinity determinations for water with pH less than 9.0

**Table 9.3**: Alkalinity Determined by Titration

| Result of Titration | Bicarbonate Alkalinity | Carbonate Alkalinity | Hydroxide Alkalinity |
|---|---|---|---|
| P = 0 | T | 0 | 0 |
| 2P = T | 0 | T | 0 |
| P = T | 0 | 0 | T |
| P<0.5T or T<2P | 0 | 2(T-P) | 2P-T |
| P>0.5T or T>2P | T-2P | 2P | 0 |

## Calculation of Bicarbonate, Carbonate and Hydroxide Alkalinity

Bicarbonate, carbonate and hydroxide alkalinity can be calculated from total alkalinity (TA) if the pH is known. These concentrations are often calculated, especially for carbonate and hydroxide ion concentrations which are generally low and difficult to measure.

$$2TA = \left(HCO_3^-\right) + 2\left(CO_3^{2-}\right) + \left(OH^-\right)$$

TA is commonly expressed as mg/L $CaCO_3$ and must be converted to moles/L (M) in order to make the following calculations. The conversion of mg/L to M $CaCO_3$ is accomplished by use of Equation next

$$\left(TA\, moles / L\, CaCO_3\right) = \frac{mg / L\, CaCO_3}{100{,}000\, mg / mole}$$

Rearranging the second disassociation constant for carbonic acid in terms of

$\left(CO_3^{2-}\right) yields$

$$K_2 = \frac{\left(H^+\right)\left(CO_3^{2-}\right)}{\left(HCO_3^-\right)} = 10^{-10.3}$$

$$\left(CO_3^{2-}\right) = \frac{K_2\left(HCO_3^-\right)}{\left(H^+\right)}$$

Substitution of equation for $\left(CO_3^{2-}\right)$ in equation above and expressing ($OH^-$) in terms of $\frac{K_w}{\left(H^+\right)}$ based on equations.

$$\left(OH^-\right) = \frac{K_w}{\left(H^+\right)} = \frac{10^{-14}}{\left(H^+\right)}$$

$$2TA = \left(HCO_3^-\right) + 2\left(\frac{K_2\left(HCO_3^-\right)}{(H)}\right) + \frac{K_w}{\left(H^+\right)}$$

Rearranging equation to solve for $HCO_3^-$ is carried out as

$$2TA - \frac{K_w}{\left(H^+\right)} = \left(HCO_3^-\right)\left(1 + \frac{2K_2}{H^+}\right)$$

$$\left(HCO_3^-\right) = \frac{2TA - 10^{Kw+pH}}{1 + 2K_2 10^{pH}}$$

Once $\left(HCO_3^-\right)$ has been calculated (or determined by titration), this concentration can be substituted into above equation and $\left(CO_3^{2-}\right)$ can be calculated. To convert these molar alkalinity concentrations to mg/L, the molar concentrations must be multiplied by the molecular weight of $CaCO_3$ (100,000 mg/mole) and in the case of $\left(HCO_3^-\right)$ and $\left(CO_3^{2-}\right)$ a factor of ½ is required because 1 mole of $CaCO_3$ is equivalent to 2 moles of $HCO_3^-$ and 2 moles of $OH^-$, whereas 1 mole of $CO_3^{2-}$ is equivalent to 1 mole of ($CaCO_3$ )

The equations necessary for calculation of bicarbonate, carbonate and hydroxide alkalinities as $CaCO_3$ mg/L are:

$$\text{Bicarbonate alkalinity (CaCO}_3\text{ mg/L)} = 50{,}000\left(\frac{2TA - 10^{Kw+pH}}{1 + 2K_2 10^{pH}}\right)$$

$$\text{Carbonate alkalinity (CaCO}_3\text{ mg/L)} = 100{,}000\ \mathrm{K_2(HCO_3^-)(10^{pH})}$$

$$\text{Hydroxide alkalinity (CaCO}_3\text{ mg/L)} = 50{,}000\left(10^{Kw+pH}\right)$$

**Carbonate Species**

The carbonate species ($H_2CO_3$), $\left(HCO_3^-\right)$ and $\left(CO_3^{2-}\right)$ concentrations can be calculated if the total dissolved inorganic carbonate (DIC) concentration and pH are known. (DIC) is often expressed as $HCO_3^-$ because it is the dominant ion at typical pH values. Figure shows the concentrations for the carbonate species for an aqueous solution with a typical DIC concentration of $10^{-3}$ M with respect to the pH of the solution. The affect of temperature is also shown. The pertinent equilibrium reactions are the two disassociation constants for $H_2CO_3$.

$$K_1 = \frac{(H^+)(HCO_3^-)}{(H_2CO_3)} = 10^{-6.4}$$

Because one mole of $CO_3^{2-}$ neutralizes 2 moles of $H^+$ the molar concentration of $CO_3^{2-}$ is multiplied by 2.

$$DIC = (H_2CO_3) + (HCO_3^-) + 2(CO_3^{2-})$$

By rearranging $K_1$ (eq. 16) to express $(H_2CO_3)$ in terms of $(HCO_3^-)$ and by rearranging $K_2$ (eq. 7) to express $(CO_3^{2-})$ in terms of $(HCO_3^-)$ and then substituting these values in equation 17 the following equation is obtained.

$$(DIC) = \left[(H^+)(HCO_3^-)(K_1^-) + (HCO_3^-) + \left[(H^+)\ (HCO_3^-)(K_2)\right]\right]$$

Factoring out $(HCO_3^-)$ and rearranging above equation to solve for $(HCO_3^-)$ and expressing $(H^+)$ as pH (negative exponent base $\log_{10}$) yields the following:

$$(HCO_3^-) = \frac{(DIC)}{\left[(10^{-pH-K_1}) + (10^{pH-K_2}) + 1\right]}$$

Because $K_1$, $K_2$ and $K_w$ are temperature dependent, the concentrations of the carbonate species must be determine using the appropriate values for a specific temperature. Using carbonate mass balance and acid equilibrium expressions, the fractional amounts of all carbonate species can be found as a function of $[H^+]$

$$\alpha H_2CO_3 = \frac{[H^+]^2}{[H^+]^2 + [H^+]K_{a1} + K_{a1}K_{a2}}$$

$$\alpha HCO_3^- = \frac{[H^+]K_{a1}}{[H^+]^2 + [H^+]K_{a1} + K_{a1}K_{a2}}$$

$$\alpha CO_3^{2-} = \frac{K_{a1}K_{a2}}{[H^+]^2 + [H^+]K_{a1} + K_{a1}K_{a2}}$$

The *pH* dependent fractions are

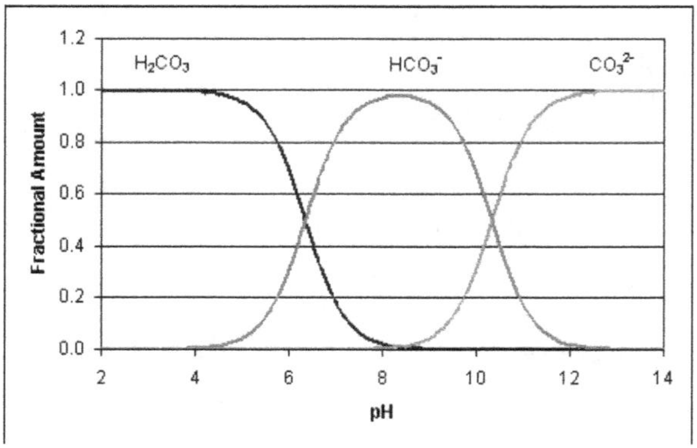


Of interest is the total carbonate as a function of p*H*.

Remember that $[H_2CO_3]$ is "fixed" by the atmosphere in an open equilibrium system and

$$\alpha H_2CO_3 = \frac{[H_2CO_3]}{\underline{F}_{H_2CO_3}} \qquad \underline{F}_{H_2CO_3} = \frac{[H_2CO_3]}{\alpha H_2CO_3}$$

Total carbonate increases with p*H*.

## 9.3 Carbonic Acid Equilibrium

### 9.3.1 Carbonic Acid

The main factor controlling carbonate solubility in nature is the amount (partial pressure) of $CO_2$ that forms $H_2CO_3$. The partial pressure for $CO_2$ in the atmosphere is 0.0003 or 0.03% of the atmospheric pressure at the surface of Earth (i.e., 1 bar or atmosphere). Partial pressure is equal to moles per liter; thus, there is $10^{-5.0}$ M $H_2CO_3$ present in water in contact with the atmosphere.

$$CO_2 + H_2O = H_2CO_3$$

$$K = \frac{(H_2CO_3)}{(CO_2)(H_2O)} = \frac{(H_2CO_3)}{(0.0003)(1)} = 10^{-1.5}$$

$$H_2CO_3 = 3\times10^{-4}\left(10^{-1.5}\right) = 10-5.0M$$

In most natural environments the appropriate equation for calcite solubility is shown i.e., $CO_2$ is present.

$$CaCO_3 + H_2CO_3 = Ca^{2+} + 2HCO_3^-$$

Above equation takes into account presence of $CO_2$ at Earth's surface which forms carbonic acid that readily dissolves $CaCO_3$.

$$K = \frac{(Ca^{2+})(HCO_3^-)^2}{(H_2CO_3)} = 10^{-4.4}$$

$(HCO_3^-) = 2\ (Ca^{2+})$ from inspection of above equation, $K = 10^{-4.4}$

$$= \frac{(Ca^{2+})(2Ca^{2+})^2}{H_2CO_3}$$

$$= \frac{4(Ca^{2+})}{10^{-5.0}}$$

The $Ca^{2+}$ concentration calculated from above equation 32 is $4.7 \times 10^{-4}$ M (equivalent to 18.8 mg/L). This is the solubility of calcite at surface conditions, i.e., 25 °C and 1 atmosphere pressure, which provides a "minimum" concentration with atmospheric carbonic acid. More $CO_2$ can be dissolved in cold water and the $CO_2$ partial in soil can be rather high leading to increased calcite solubility. Decaying vegetation in soil can have $CO_2$ partial pressure in excess of 0.1 atmospheres which can significantly raise $H_2CO_3$ concentrations of infiltrating meteoric water. The solubility of calcite in water in contact with 0.1 atmosphere of $CO_2$ can be calculated as shown, producing a calcite solubility of $3.2 \times 10^{-3}$ M (equivalent to 128 mg/L $Ca^{2+}$). This value is not an absolute maximum but does give an approximate upper limit for $Ca^{2+}$ concentrations from limestone near the surface of Earth. Other acids, e.g., sulfuric from the oxidation of sulfide minerals such as pyrite ($FeS_2$), can also have a significant affect on calcite solubility locally.

**Cave Formation**

Carbonic acid-rich water forms limestone caves, which are the most common type of caves. When the water table is high, carbonic acid-rich water dissolves the limestone (calcite). Later when the water table drops, a "void" filled with air is formed and a large void is called a cave. Smaller amounts of water rich in $Ca^{2+}$ and $HCO_3^-$ may continue to flow through the void. When these waters enter the void, $CO_2$ partial pressure decreases and $CO_2$ is released. This degassing of $CO_2$ drives the reactions in equations 27 and 30 to the left, leading to the precipitation of calcite and formation of stalagmites, stalactites and other cave features. Although evaporation may play a small role in the deposition of calcite, it is the loss of $CO_2$ that is the most important factor.

**Hardness**

Water hardness is a measure of how difficult it is for soap to lather, i.e., most of the divalent ions must be precipitated with soap before soap can produce suds. The most abundant divalent ions in nature are $Ca^{2+}$ and $Mg^{2+}$, although $Fe^{2+}$ and other divalent ions can be important in some waters. The major source of $Ca^{2+}$ and $Mg^{2+}$ is carbonate minerals; therefore, hardness is expressed as mg/L $CaCO_3$. The range in hardness values with descriptive terms is shown in Table 9.4.

**Table 9.4**: Hardness Ranges. Units are in mg/L of calcium carbonate. After Hem (1985).

| | |
|---|---|
| 0-60 | Soft |
| 51 – 120 | Moderately Hard |
| 121-180 | Hard |
| >180 | Very Hard |

In the past, hardness was determined by a titration analysis of divalent ions. Today hardness is calculated by converting $Ca^{2+}$ and $Mg^{2+}$ molar concentrations to mg/L $CaCO_3$ by multiplying the mg/L concentrations of the two ions by the inverse of the equivalent weights of the ions and summing the two concentrations.

$$\text{Hardness as mg/L CaCO}_3 = 2.497(\text{Ca}^{2+}\, mg/L) + 4.118\ (\text{Mg}^{2+}\, mg/L)$$

Some scientists consider hardness more of a water quality (use) factor that relates to scaling in hot water heaters and in industrial setting, rather than an important geochemical factor. However, comparison of hardness and alkalinity can yield helpful water chemistry information. If hardness exceeds alkalinity, the excess is termed "non-carbonate" hardness, meaning that there are non-carbonate mineral sources of $Ca^{2+}$ and $Mg^{2+}$ ions. The remainder of the hardness is considered to be "carbonate" hardness and derived from carbonate minerals. In this discussion it is assumed that there is "no" hydroxide alkalinity. If alkalinity exceeds hardness, there must be carbonate mineral sources that do not contain $Ca^{2+}$ and/or $Mg^{2+}$ e.g., $NaHCO_3$ or cation exchange has occurred with clay minerals.

### Dolomite

Dolomite has a very small $K_{sp}$ with values ranging from $10^{-17}$ to $10^{-19}$ The $K_{sp}$ is too small to produce large thicknesses of dolomite and furthermore there is no true dolomite being precipitated today. Some "proto-dolomite" is precipitated in restricted environments, e.g., evaporite lagoons. Dolostone characteristically has poor preservation of fossils, is coarse grained and commonly has cavities/pore spaces, which all indicate a replacement of limestone as the origin of the rock. In addition, it is known that saline water, high pH, low $Ca^{2+}/Mg^{2+}$ ratio and elevated temperatures all favor dolostone formation. Based on these factors, various marine and deep water reflux models have been proposed, as well as mixing models for fresh water and sea water but no "perfect" model has been developed for the formation of dolomite. The dolomite atomic structure is highly ordered which may affect the amount of time necessary to develop the structure. At any rate the formation and dissolution of dolomite is complex and not well understood in comparison with other carbonate minerals. Dolomite undergoes incongruent dissolution with $Mg^{2+}$ dissolving before $Ca^{2+}$ .

### Carbonic Acid ($H_2CO_3$)

Hypothetical acid of carbon dioxide and water exists only in form of its salts (carbonates), acid salts (hydrogen carbonates), amines (carbamic acid), and acid chlorides (carbonyl chloride). If we add a weak acid such as carbonic acid in water it weakly dissociates to form bicarbonate and hydrogen ions . Now water already has some hydrogen ions, having a ph=7. So now indirectly we added hydrogen ions and base (bicarbonate ) ions to water which already has a hydrogen ion concentration. The resulting solution is acidic . This means that in the addition of equimolar hydrogen ions and bicarbonate ions( base) hydrogen ions dominate.

The atmospheric gas carbon dioxide ($CO_2$) dissolves very easily in water. This is well known in mineral water, which often has carbon dioxide added. In the dissolution process, carbon dioxide reacts with the water molecules according to the equation below. When carbon dioxide mixes with the water it is partially converted into carbonic acid, hydrogen ions ($H^+$), bicarbonate ($HCO_3^-$), and carbonate ions ($CO_3^{2}$). Seawater can assimilate much more $CO_2$ than fresh water. The reason for this is that bicarbonate and carbonate ions have been perpetually discharged into the sea over aeons. The carbonate reacts with $CO_2$ to form bicarbonate, which leads to a further uptake of $CO_2$ and a decline of the $CO_3^{2-}$ concentration in the ocean. All of the $CO_2$-derived chemical species in the water together, i.e. carbon dioxide, carbonic acid, bicarbonate and carbonate ions, are referred to as dissolved inorganic carbon (DIC). This carbonic acid-carbonate equilibrium determines the amount of free protons in the seawater and thus the pH value.

$$CO_{2+H2}O \leftrightarrow H_2CO_{3 \leftrightarrow H}{}^{++} HCO_3^- \leftrightarrow 2\ H^{++} CO_3^{2-}$$

Hence a reaction of carbon dioxide in seawater shows that initially carbon dioxide reacts with water to form carbonic acid. This will further react with carbonate ions and forms bicarbonate. Over long term, ocean acidification decrease in concentration of carbonate ions in seawater. A 50 per cent decline of the levels is predicted, for example, if there is a drop in pH levels of 0.4 units. This would be fatal. Because carbonate ions together with calcium ions ($CaCO_3$) form the basic building blocks of carbonate skeletons and shells, this decline would have a direct effect on the ability of many marine organisms to produce biogenic carbonate. In extreme cases this can even lead to the dissolution of existing carbonate shells, skeletons and other structures.

## 9.4 Ionization Equilibra

### 9.4.1 Photoionization Equilibrium

At every point in a optically thin, pure hydrogen medium of neutral density $n_{HI}$, the photoionization rate per unit volume is

$$n_{HI} \int_{vL}^{\infty} \frac{4\pi J_v \sigma_H(v)}{hpv} dv,$$

where

$Jv$ = mean intensity of ionizing radiation

$h_P$ = Planck constant

The photoionization cross-section for hydrogen in the ground state by photons with energy $h_p v$ (above the threshold $h_p v_L = 13.6$ eV) can be usefully approximated by

$$\sigma_H(v) = \sigma(v / v_L)^{-3}, \qquad \sigma_L = 6.3 \times 10^{-18} cm^2.$$

At equilibrium, this is balanced by the rate of radiative recombination's $p + e \to H + \gamma$ per unit volume,

$$n_\varepsilon n_p \alpha_A(T),$$

where

$n_e$ and $n_p$ = number densities of electrons and protons

$\alpha_A \Sigma < \sigma_n v_e >$ = radiative recombination coefficient which is the product of electron capture cross-section $\sigma_n$ and electron velocity $v_e$, averaged over a thermal distribution and summed over all atomic levels $n$.

At the commonly encountered gas temperature of $10^4$ K, $\alpha_A = 4.2 \times 10^{-13}$ cm$^3$ s$^{-1}$. Consider, as an illustrative example, a point in an intergalactic H II region at $z = 6$, with density $\bar{n}_H = (1.6 \times 10^{-7}$ cm$^{-3})$ $(1 + z)^3 = 5.5 \times 10^{-5}$ cm$^{-3}$. The H II region surrounds a putative quasar with specific luminosity $Lv = 10^{30} (v_L / v)^2$ ergs s$^{-1}$ Hz$^{-1}$, and the point in question is at a distance of $r = 3$ Mpc from the quasar. To a first approximation, the mean intensity is simply the radiation emitted by the quasar reduced by geometrical dilution,

$$4\pi J_v = \frac{L_v}{4\pi r^2}$$

For photoionization timescale:

$$t_{ion} = \left( \int_{vL}^{\infty} \frac{4\pi v \sigma(v)}{hpv} dv \right)^{-1} = 5 \times 10^{12} s,$$

For recombination timescale:

$$t_{rec} = \frac{1}{n_e \alpha_A} = 5 \times 10^{16} s \frac{\bar{n}H}{n_e}.$$

As in photoionization equilibrium $n_{HI} / t_{ion} = n_p / t_{rec}$, these values imply $n_{HI} / n_p$ □ $10^{-4}$, that is, hydrogen is very nearly completely ionized.

A source radiating ultraviolet photons at a finite rate cannot ionize an infinite region of space, and therefore there must be an outer edge to the ionized volume. One fundamental characteristic of the problem is the very small value of the mean free path for an ionizing photon if the hydrogen is neutral, $(\sigma_L n_H)^{-1} = 0.9$ kpc at threshold, much smaller than the radius of the ionized region.

If source spectrum is steep enough that small amount of energy is carried out by more penetrating, soft X-ray photons, then nearly completely ionized H II region separated from the outer neutral IGM by thin transition layer or ionization-front takes place. The in-homogeneity of the IGM is of primary importance for understanding the ionization history of the Universe, as denser gas recombines faster and is therefore ionized at later times than the tenuous gas in underdense regions. An approximate way to study the effect of in-homogeneity is to write the rate of recombination's as

$$\langle n_e n_p \rangle \alpha_A(T) = C \langle n_\varepsilon \rangle^2 \alpha_A(T)$$

where = brackets are space average of product of proton and electron number densities

$C > 1$ = degree of clumpiness of IGM

If ionized gas with electron density $n_e$ density filled uniformly a fraction $1/C$ of the available volume, the rest being empty space, the mean square density would be

$$< n_e^2 > = n_e^2 / C = < n_e >^2 C.$$

The IGM is completely reionized when one ionizing photon has been emitted for each H atom by the radiation sources, and when the rate of emission of UV photons per unit (comoving) volume balances the radiative recombination rate, so that hydrogen atoms are photoionized faster than they can recombine. The complete reionization of the Universe manifests itself in the absence of an absorption trough in the spectra of galaxies and quasars at high redshifts. If the IGM along the line-of-sight to a distant source were neutral, the resonant scattering at the wavelength of the Ly$\alpha$ ($2p$ -> $1s$; $h_p \nu \alpha = 10.2$ eV) transition of atomic hydrogen would remove all photons blueward of Ly$\alpha$ off the line-of-sight. For any reasonable density of the IGM, the scattering optical depth is so large that detectable absorption will be produced by relatively small column (or surface) densities of intergalactic neutral hydrogen.

**Radiation Fields**

The E and H fields which are set to transfer energy by space are radiation field which are responsible for electromagnetic radiation from antenna. Radiation field decreases as distance from antenna increases as the decrease is linear, radiation field reaches great distances from the antenna.

In case of half-wave antenna, radiation actually takes place which is one having electrical length equal to half wavelength of transmitted signal. If a transmitter is operating at 30 megahertz and an half-wave antenna is applied with transmitter, then antenna's electrical length would have to be at least 16 feet long.

When power is delivered to the half-wave antenna, both an induction field and a radiation field are set up by the fluctuating energy. At the antenna, the intensities of these fields are proportional to the amount of power delivered to the antenna from a source such as a transmitter. At a short distance from the antenna and beyond, only the radiation field exists. This radiation field is made up of an electric component and a magnetic component at right angles to each other in space and varying together in intensity. With a high-frequency generator connected to antenna, the induction field is produced, but the generator potential reverses before electrostatic field has had time to collapse. The reversed generator potential neutralizes the remaining antenna charges, leaving a resultant E field in space. Figure shows an E field detaching itself from antenna where view A shows voltage is maximum and electric field having maximum intensity.

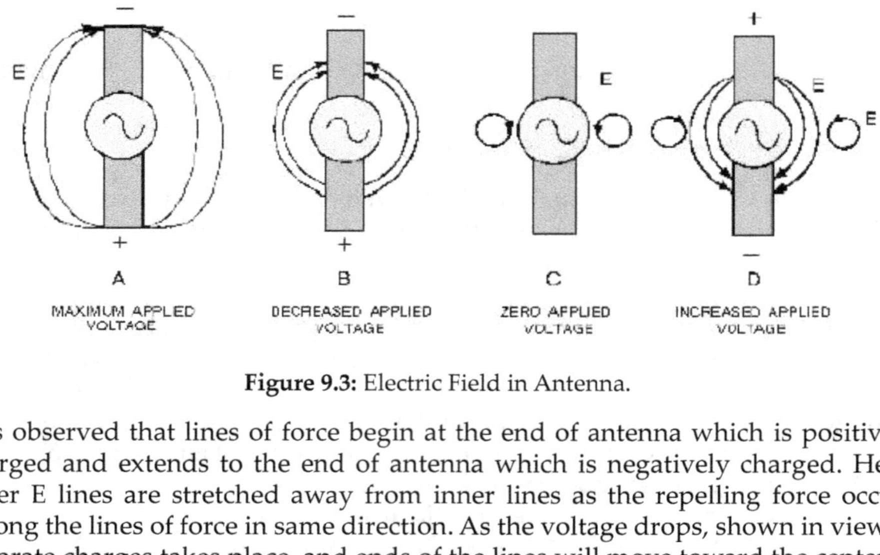


**Figure 9.3:** Electric Field in Antenna.

It is observed that lines of force begin at the end of antenna which is positively charged and extends to the end of antenna which is negatively charged. Here, outer E lines are stretched away from inner lines as the repelling force occurs among the lines of force in same direction. As the voltage drops, shown in view B, separate charges takes place, and ends of the lines will move toward the center of antenna. Further, since lines of force are in same direction tends to repel each other where the centers of the lines are held outside.

As the voltage approaches zero (view B), some of the lines collapse back into the antenna. At the same time, the ends of other lines begin to come together to form a complete loop. Notice the direction of these lines of force next to the antenna in view C. At this point the voltage on the antenna is zero. As the charge starts to build up in the opposite direction (view D), electric lines of force again begin at the positive end of the antenna and stretch to the negative end of the antenna. These lines of force, being in the same direction as the sides of the closed loops next to the antenna, repel the closed loops and force them out into space at the speed of light. As these loops travel through space, they generate a magnetic field in phase with them.

Since each successive E field is generated with a polarity that is opposite the preceding E field (that is, the lines of force are opposite), an oscillating electric field is produced along the path of travel. When an electric field oscillates, a magnetic field having an intensity that varies directly with that of the E field is produced. The variations in magnetic field intensity, in turn, produce another E field. Thus, the two varying fields sustain each other, resulting in electromagnetic wave propagation. During this radiation process, the E and H fields are in phase in time but physically displaced 90 degrees in space. Thus, the varying magnetic field produces a varying electric field; and the varying electric field, in turn, sustains the varying magnetic field. Each field supports the other, and neither can be propagated by itself.

# 10 pH and Equilibrium

## 10.1 Introduction

When the forward and reverse rates are equal, there is no further change in the concentrations of reactants and products and the reaction is said to have reached equilibrium. If the forward reaction is intrinsically fact and the reverse reaction slow, equilibrium will occur when the concentration of reactants is very low and that of products is very high. Equilibrium is said to lie to the right or to favor products.

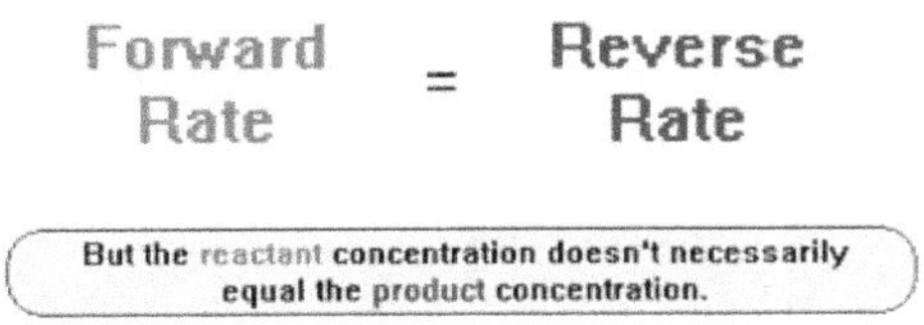


**Figure 10.1**: Equilibrium.

If the forward reaction is intrinsically slow and the reverse reaction fact equilibrium will occur when the concentration of reactants is still very high and that of products is very low. Equilibrium is said to lie to the left of to favor reactants.

Although, at equilibrium, there is no further change in concentrations, both the forward and reverse reactions are still taking place, the system is not static and so the equilibrium is called a dynamic equilibrium.

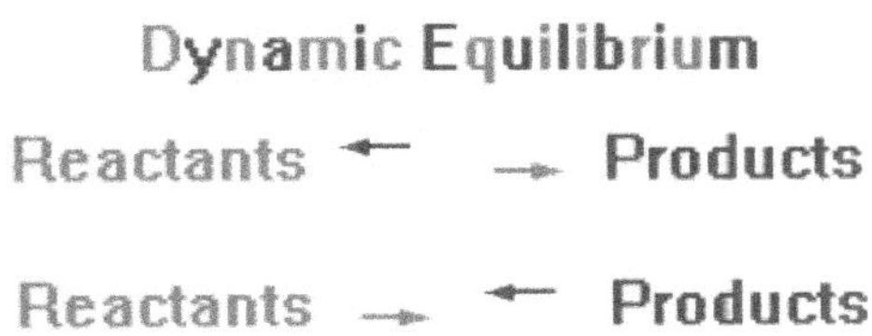


**Figure 10.2**

On changing the concentration of reactant or product molecules for disturbing an equilibrium, where both the number of molecules and volume of solution changes so that the concentrations and equilibrium will remain unaffected can be done when only reactant molecules are added, where rate of forward reaction increases there unaffecting rate of reverse reaction as the concentration of product molecules are not changed.

If the reactant are added, then there will be change in total volume occupied by reactants and products by diluting them that lead to lower concentration of product molecules. This has the effect of slowing the reverse reaction a little. Then concentration of reactants begins again to decrease, while concentration of products begins to increase in such case, the equilibrium will shifts to the right. It is observed that response of system countered the change as imposed by adding reactant molecules to mixture where equilibrium shifts towards right.

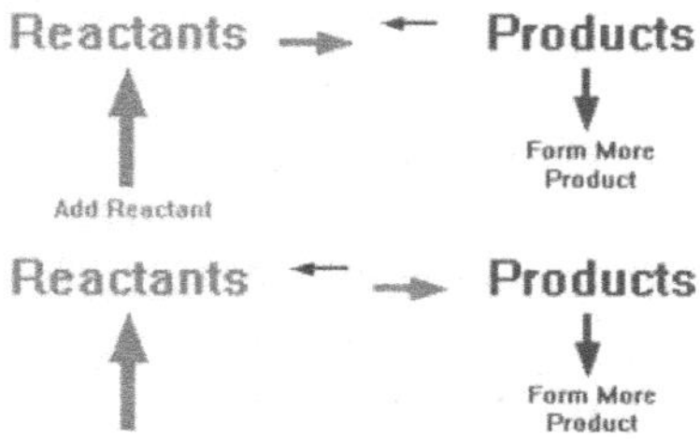


**Figure 10.3**: Dynamic Equilibrium.

It is noted that when product concentration increases, reaction will shifts to the left thereby decreasing product concentration and increasing reactant concentration. Such principle showing any change imposed on system at equilibrium resulting in equilibrium to shift in direction that counteracts the change is known as Le Chatelier's Principle which applies to all dynamic equilibrium. Hence an equilibrium will shift to right when concentration of reactant increases or concentration of product decreases.

This will shift towards left when concentration of reactant is decreased or that of product is increased. If a change is imposed which equilibrium cannot counteract by shifting in one direction or the other, the equilibrium will not be affected by the change.

In case of acid-base equilibria, pH measures $H^+$ ion concentration and $OH^-$ ion concentration. It is known that pH = -log $[H^+]$, where as concentration of $H^+$ increases, pH decreases. Since the concentrations of $H^+$ and $OH^-$ are inversely proportional, so as one increases, other will decreases.

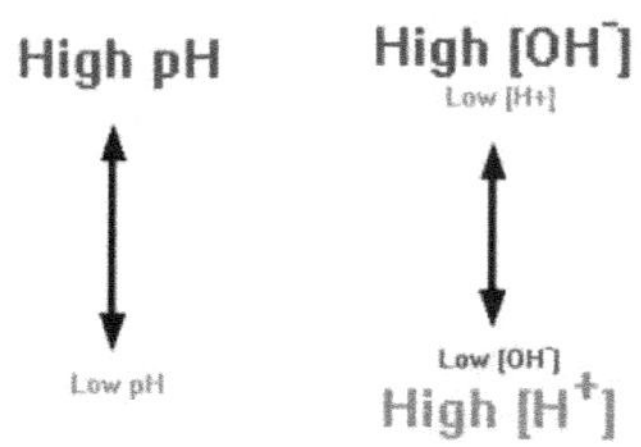


**Figure 10.4**: Variation of pH

In acid-base equilibrium, decreasing the pH amounts increases $H^+$ concentration that results in equilibrium to shift to the left. The opposite occurs when pH increases and $H^+$ concentration decreases using a base. The $H^+$ ions tends to be removed from solution, but $OH^-$ ions in base react with $H^+$ ions to form water that effectively takes out of acid-base equilibrium reaction which further lower the concentration of $H^+$ ions.

### 10.1.1 pH of Substances

Acidic solutions have pH values less than 7 while basic solutions have pH values more than 7. It is observed that more acidic the solution is, lower will be its pH value while the more basic the solution is, higher will be its pH value. The corresponding $H^+$ and $OH^-$ concentrations are shown in units of molarity where decrease of 1 pH unit corresponds to 10 fold increase in [H+] while increase of 1pH unit for basic solution corresponds to 10 fold increase in $[OH^-]$.

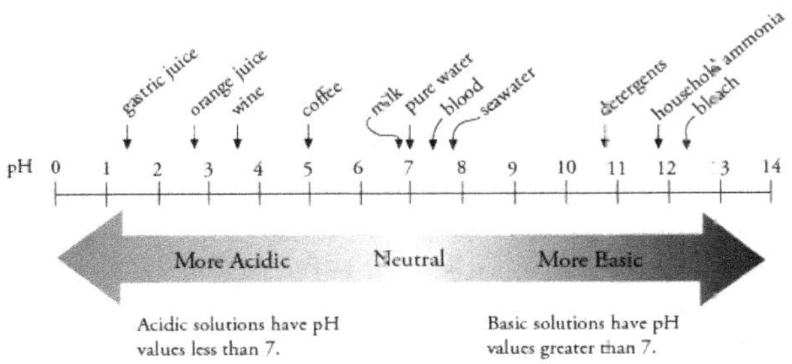


**Figure 10.5**: pH scale

The pH, $[H^+]$, and $[OH^-]$ of some common solutions are shown in figure where it is observed that gastric juice, which is our stomach will have a pH of 1.4 while orange juice will have a pH of 2.8. So it is noted that gastric juice will be more than 10 times more concentrated in $H^+$ as compared to orange juice. The pH difference of about 4 between household ammonia solutions (pH about 11.9) and milk (pH about 6.9) shows that household ammonia has about ten thousand ($10^4$) times the hydroxide concentration of milk.

**Example** If the $H^+$ concentration of 2.9 × 10-3 NaOH solution is 5.1 ×10-12 M $H^+$ then what will be its pH?

**Solution**

$$pH = -\log[H^+]$$

$$= -\log(7.5 \times 10^{-12})$$

$$= 11.29$$

Now converting from pH to $[H^+]$ and $[OH^-]$ using equations

$$[H^+] = 10^{-pH}$$

$$\left[OH^-\right]=\frac{K_w}{\left[H^+\right]}$$

## 10.2 Characteristics of pH

pH is one of the most commonly determined parameters of aqueous (and non-aqueous) solutions because many chemical processes are dependent on pH. Altering pH can often significantly alter the rate of chemical reactions. The solubility and bioavailability of many pharmaceutically active compounds is also determined by solution pH. Furthermore, in chromatography the retention and selectivity of ionizable analytes can be manipulated by changing the eluent pH.

pH is a unit of measure that describes the acidity or basicity of a solution and is measured on a scale of 0 to 14, pH 0 being very 'acidic', pH 14 being very 'basic'. The term is derived from 'p' the mathematical symbol for the negative logarithm, and 'H' the chemical symbol for hydrogen.

If the pH of a solution is not within the correct range the plant will not have the ability to absorb some of the essential elements required for proper plant growth. All plants have a particular pH range, which will produce healthy growth, and this level will vary from plant to plant, but most plants prefer a slightly acidic growing environment (5.8 to 6.2), although most plants can survive in an environment with pH values between 5.0 and 7.0.

For our body pH is one of the most important elements affecting health status and function. It reflects the events occurring in our body. Many conditions and diseases have this one basic element in common, out of range pH levels. In fact, pH influences every physiologic activity in your body, including metabolism, pain, and diseases. The speed of all biological and electrical reactions is under pH control. For example, when the body environment is more acidic, electrical resistance lowers, hence electricity moves faster. Conversely, the more alkaline the body is the slower electricity in the body. Also, the more acidic the blood, the quicker our body wants to get rid of the acidity, so a person's breaths are quicker and shallower. If a person's pH sways too far out of range the cells can be poisoned by their own toxins and die.

Hard water is characterized by high levels of bicarbonates and it makes itself known by depositing lime scale in kettles and by reducing the lathering ability of soap. Hard water will usually have a high pH but not necessarily. What will distinguish hard water is that it will take much more acid to reduce the pH than with an equivalent sample of softer water. This is because the bicarbonates have to be neutralized and this uses up quite a lot of acid.

## 10.3 Acid/Base Calculations

### 10.3.1 Strong Acids and Bases

It is known that calculations using strong acids and bases are clear as such species dissociate completely in water, so, [strong acid] = $[H^+]$. In case for strong bases, study groups I and II having hydroxide ($OH^-$) and oxide ($O^{2-}$) salts. NaOH provides one mole of $OH^-$ per mole of salt, but $Ca(OH)_2$ will give two moles of $OH^-$ per mole of salt. After having the $[H^+]$ or $[OH^-]$, it is easy to change these into pH or pOH quantities.

### 10.3.2 Weak Acids and Bases

Weak acids and base are challenging as they undergo incomplete dissociation, and exist as systems at equilibrium. For this, the correct approach is most systematic approach which includes:

1. Writing balanced equation and expression for *K*. Comparing these with information that you have been given, for example, initial concentrations, equilibrium concentrations, the value of *K*, etc which helps in deciding what to find out.
2. Define **x** as unknown concentration which changes during the reaction. It can be shown by the stoichiometry of reaction.
3. Prepare a reaction table which incorporates the unknown.
4. Make assumptions that simplifies the calculations.
   a. As values of *K* are small, so very little of initial acid or base gets dissociated. While doing preliminary calculations, it is assume that $[HA]_i \approx [HA]_{eq}$.
   b. Assume $[H^+]$ and $[OH^-]$ in a solution due to autoionization of water is negligible as it becomes important at very low ($< 10^{-6}$ M) concentrations of acids or bases, when water act as main source of $H^+$ and $OH^-$.

**Example**

Finding the $K_a$ of a weak acid from the **pH** of its solution.

A **0.10M** solution of **formic acid**, HCOOH, has a **pH = 2.38** at 25°C. Calculate the $K_a$ of formic acid.

$$HCOOH\ (aq) \leftrightarrow HCOO^- + H^+ \quad K_a = [HCOO^-]_{eq}\ [H^+]_{eq}$$

$$[HCOOH]_{eq}$$

It is known that equilibrium concentration of $H^+$ as pH of solution is given as $[H^+]_{eq} = 10^{-2.38} = 4.2 \times 10^{-3}$ M

As initial concentration of HCOOH, so losing **x** moles of this as acid undergoes dissociation to form $HCOO^-$ and $H^+$.

| | [HCOOH](M) | $[H^+]$(M) | $[HCOO^-]$(M) |
|---|---|---|---|
| Initial | 0.10 | 0.00 | 0.00 |

Change -x +x +x

Equilibrium 0.10 -x x x

$$K_a = \frac{(x)(x)}{(0.10 - x)}$$

If all of $H^+$ comes from acid and none from water, then $[H^+]_{eq} = 4.17 \times 10^{-3}$ which is much higher than $1.0 \times 10^{-7}$ M $H^+$ from water, so the concentration of HCOOH will change very little, from 0.10 to $0.10 - 4.17 \times 10^{-3}$. The change in concentration can be ignored if it is less than **5%** of the original concentration. 0.10 M x 5% = $5.0 \times 10^{-3}$, so the change in [HCOOH] in this problem can be ignored.

$$K_a = \frac{(x)(x)}{(0.10)} = \frac{(4.2 \times 10^{-3})^2}{(0.10)} = 1.8 \times 10^{-4}$$

Now calculate percent ionization as it shows relative number of acid molecules which dissociate, so calculating as

$$\frac{[H^+]_{eq} \times 100}{[HCOOH]_i}$$

Here, percent ionization = $\{(4.2 \times 10^{-3})/ (0.10)\} \times 100 = 4.2\%$ which shows that small of acid dissociates under such conditions. Hence for strong acids and bases, the percent ionization is 100%.

**Example**

Finding the **pH** of a solution, given the $K_a$ of a weak acid. Calculating pH of 0.10 M HF solution where $K_a = 6.8 \times 10^{-4}$ at 25°C.

**Solution**

1. $HF(aq) \leftrightarrow F^-(aq) + H^+(aq)$ $K_a = 6.8 \times 10^{-4} = \frac{[F^-]_{eq} [H^+]_{eq}}{[HF]_{eq}}$
2. Initial [HF] and $K_a$ are known
3.

| | [HF] (M) | $[H^+]$(M) | $[F^-]$(M) |
|---|---|---|---|
| Initial | 0.10 | 0.00 | 0.00 |
| Change | -x | +x | +x |
| Equilibrium | 0.10 - x | x | x |

4. Assume that $[H^+]$ comes only from acid, and that decrease in initial concentration of HF which is negligible.

$$K_a = 6.8 \times 10^{-4} = \frac{(x)(x)}{(0.10)} \quad x = 8.2 \times 10^{-3}$$

Now we need to check if this is less than 5% of the original concentration, to see if our simplification is allowed.

$$\frac{8.2\times10^{-3}}{0.10}\times100 = 8.2\%$$

$$K_a = 6.8 \text{ x } 10^{-4} = \frac{(x)(x)}{(0.10)}$$

Now expressing in quadratic equation, and solving

$$x^2 + 6.8 \times 10^{-4} x - 6.8 \times 10^{-5} = 0$$

$$x = -b \pm \sqrt{b^2 - 4 ac} /2 a$$

$$x = 7.9 \times 10^{-3} \text{ M} = [H^+]$$

$$pH = - \log (7.9 \times 10^{-3}) = 2.10$$

Now again calculate percent ionization:

$(7.9 \text{ x } 10^{-3})/(9.10) \text{ x } 100 =$ **7.9 %**

So, a 0.10 M solution of HF will ionize to a greater extent (7.9%) than a 0.10 M solution of formic acid, HCOOH (4.2%). It is a stronger acid.

**Example**

What is the pH of a **0.010 M** solution of HF?

**Solution**

Similar to above example,

$$K_a = \frac{(x)(x)}{(0.010-x)} = 6.8\times10^{-4}$$

Now solving for x using quadratic equation gives

$$x = [H^+] = 2.3 \times 10^{-3} \text{ M}$$

Now the pH = 2.64 which is less acidic as concentration of $H^+$ in more dilute system is lower.

Now percent ionization:

$$= ([H^+] / [HF]) \times 100$$

$$= \{(2.3 \text{ x } 10^{-3})/ (0.010)\} \times 100$$

$$= 23.0\%$$

As $[H^+]$ is lower, so the percent ionization is much higher in dilute acid. It is observed that lowering the volume of gaseous system will shift the equilibrium toward the side having lesser moles of gas. In aqueous systems, similar effect of concentration is seen while in acid dissociation, $M^2$ in products and M in reactants will be observed. Decreasing the concentration of all of the species will make $Q < K_a$, since we are squaring the numerator (products), but not the denominator(reactant). More of the acid will ionize, if the concentration of the acid is lower.

**Example**

Calculate pH of a 0.15 M solution of $NH_3$. The $K_b$ of ammonia at 25°C is $1.8 \times 10^{-5}$.

**Solution**

1. $NH_3 \leftrightarrow NH_4^+ + OH^-$ $\quad$ $K_b = \frac{[NH_4^+]_{eq}[OH^-]_{eq}}{[NH_3]_{eq}}$

2. Now $K_b$ and $[NH_3]_I$ are known

3.

| | $[NH_3]$ | $[NH_4^+]$ | $[OH^-]$ |
|---|---|---|---|
| Initial | 0.15 | 0.00 | 0.00 |
| Change | -x | +x | +x |
| Equilibrium | 0.15 - x | x | x |

4. Assume $OH^-$ coming from autoionization of water will be small and where less ammonia will be ionizes. Now checking the assumptions

$K_b = 1.8 \times 10^{-5} = \frac{(x)(x)}{0.15}$ $\quad$ $x = 1.64 \times 10^{-3} = [OH^-]$

$$\frac{1.64 \times 10^{-3}}{0.15} \times 100 = 1\%$$

= **percent ionization** and proof that a simplifying assumption is made about x << 0.15.

This $[OH^-]$ is also much greater than $1 \times 10^{-7}$ M $OH^-$ contributed by water.

As $[OH^-]$ is known so solving for pH:

pH + pOH = 14

or $[H^+][OH^-] = 1 \times 10^{-14}$

$[OH^-] = 1.64 \times 10^{-3}$, so

$pOH = -\log(1.64 \times 10^{-3})$

= 2.79

Now 2.79 + pH = 14

pH = 11.21

or $[H^+] [1.64 \times 10^{-3}] = 1 \times 10^{-14}$

$[H^+] = 6.10 \times 10^{-12}$

pH = 11.21

**Example** Find the pH of a 0.0037 M solution of $H_2CO_3$. This acid has two protons that can be dissociated, and two $K_a$ values.

$H_2CO_3\ (aq) \leftrightarrow H^+\ (aq) + HCO_3^-\ (aq)\ K_{a1} = 4.3 \times 10^{-7}$

$HCO_3^-\ (aq) \leftrightarrow H^+\ (aq) + CO_3^-\ (aq)\ K_{a2} = 5.6 \times 10^{-11}$

The first acid, $H_2CO_3$ is a much stronger acid, judging from its higher value of $K_a$. First calculate the pH and equilibrium concentrations after the first proton dissociates.

| | $[H_2CO_3]$ (M) | $[H^+]$ (M) | $[HCO_3^-]$ (M) |
|---|---|---|---|
| Initial | 0.0037 | 0.00 | 0.00 |
| Change | -x | +x | +x |
| Equilibrium | 0.0037 - x | x | x |

$$4.3\times10^{-7} = \frac{x^2}{0.0037 - x} \quad x = 4.0\times10^{-5}$$

To test the assumption that × is <<< 0.0037, $\frac{4.0\times10^{-5}}{3.7\times10^{-3}}$ × 100 = 1.1%, so the assumption is O.K.

$3.7 \times 10^{-3}$

So, $[H_2CO_3] = 0.0037$ M, as dissociation was so small.

$[HCO_3^-] = [H^+] = 4.0 \times 10^{-5}$M after the first dissociation reaction.

Now use such concentrations as initial concentrations in second acidic reaction.

$HCO_3^-\ (aq)$ D $H^+\ (aq) + CO_3^{2-}\ (aq)\ K_a = 5.6 \times 10^{-11}$

| | $[HCO_3^-]$ (M) | $[H^+]$ (M) | $[CO_3^{2-}]$ (M) |
|---|---|---|---|
| Initial | $4.0 \times 10^{-5}$ | $4.0 \times 10^{-5}$ | 0.00 |
| Change | -y | +y | +y |
| Equilibrium | $4.0 \times 10^{-5}$ - y | $4.0 \times 10^{-5}$ + y | y |

$$5.6\times10^{-11} = \frac{\left(4.0\times10^{-5} + y\right)\left(y\right)}{\left(4.0\times10^{-5} - y\right)}$$

$y = 5.6 \times 10^{-11} = [CO_3^{2-}]$

To test our assumptions $\dfrac{5.6\times10^{-11}}{4.0\times10^{-5}}\times100 = 1.4\times10^{-4}\,\%$

So, the pH at equilibrium will be determined exclusively by the first step,

$[H^{=}] = 4.0 \times 10^{-5}$,

pH = 4.4

## 10.4 Determining pH of Weak Acids and Weak Bases

### 10.4.1 Weak Acids

As per Bronsted and Lowry, all acids began with $H^{+1}$ ion tend to behave as acids, but did not completely dissociate. Weak acid is that doesn't completely dissociate or dissolve in water. Such type of incomplete dissociation is shown by two-way reaction arrow between the reactant side of a chemical reaction and product side described as reversible reaction.

In an example, acetic acid is an acid described by Bronsted and Lowry as weak acid as it tends to dissociate into ions when dissolved in water

$HC_2H_3O_2(aq)\ H^{+1}(aq) + C_2H_3O_2^{-1}(aq)$

acetic acid hydrogen cation acetate anion

In the reactions, double arrow shows a reverse reaction. Other weak acids identified by Bronsted and Lowry include:

HCN (aq) (cyanic acid) $H^{+1}$ (aq) + CN-1 (aq)

$HCO_2H$ (aq) (formic acid) $H^{+1}$(aq) + $CO_2$H-1 (aq)

$H_2C_2O_4$ (aq) (oxalic acid) $H^{+1}$ (aq) + $HC_2O_4^{-1}$ (aq)

Here, oxalic acid only loses one of its H+1 cations upon dissociation, then soon receives it right back. Also, the $H^{+1}$ cation lost in differing degrees by both strong and weak acids which is a hydrogen atom that loses an electron while its nucleus contains one proton. Therefore, as we turn our discussion to weak bases, and the reactions of weak acids/weak bases, remember that:

H+1 = a proton

### 10.4.2 Weak Bases

As per Bronsted and Lowry theory, theories of acid, strong or weak will tend to lose it H+1 cation (proton) when dissolved in water. It is merely a matter of whether the H+1 is dissociated permanently (strong acids), or if there is any possibility of recombination back into the original acid form (weak acids).

Bronsted and Lowry discovered bases that fall well outside Arrhenius' original definition (namely, that all bases end in the OH-1 anion). They still supported

Arrehnius' version to define strong bases, but they found that weak bases also existed, and didn't end in the OH-1 anion.

Bronsted and Lowry observed that the weak acids, in losing their proton (the H+1 cation), however temporarily), actually gave, or donated, the proton to substances that had an attraction for just such a positive ion! Using the general idea that "opposites attract", Bronsted and Lowry found that compounds containing and beginning with a nitrogen atom ($NH_3$, $NF_2Cl$, $NBr_3$, etc) were good receivers of the proton being lost by the acid, due to the lone-pair electrons (negatively-charged, of course) on the nitrogen atom.

Now simplify the problem of weak acid by starting with general set of equations:

$$K_w = \left[H^+\right]\left[OH^+\right]$$

$$\left[H^+\right] = \left[OH^-\right] + \left[A^-\right]$$

$$K_a = \frac{\left[H^+\right]\left[A^-\right]}{[HA]}$$

$$\left[A^-\right] + [HA] = C_a$$

If the acid is not very diluted and not extremely weak above equation will take form of

$\left[H^+\right] \approx \left[A^-\right]$ that reflects stoichiometry of dissociation.

Also, remove water ion product from as $[OH^-]$ is no longer present in equation, so we left with the equations. On substituting $[H^+]$ for $[A^-]$ in above equation gives

$$K_a = \frac{\left[H^+\right]}{[HA]}$$

From fourth equation

$[HA] = C_a - \left[H^+\right]$ and inserting in equation sixth to get

$$K_a = \frac{\left[H^+\right]^2}{C_a - \left[H^+\right]}$$

or in polynomial form

$$\left[H^+\right]^2 + K_a\left[H^+\right] - K_a C_a = 0$$

This is quadratic equation for [H⁺] concentration with two roots of which only one is positive:

$$\left[H^+\right] = \frac{-K_a + \sqrt{K_a^2 + 4K_aC_a}}{2}$$

It is observed that above equation is similar with result obtained when using Initial, Change, Equilibrium method for weak acid dissociation. It is noted that for very weak acids when $K_a$ approaches 0, then numerator also approaches to zero and result of pH calculation can be higher than 7. As an example with $pK_a$=14 and $C_a$=0.01 M and putting pH=8, then the effect of neglecting water dissociation will be similar to as observed when we used equation $\left[H^+\right] \approx C_a$ for strong acids.

With rules as if one of elements in sum is less than 5% of second, it can be neglected. Many dissociation constants are known with 1% accuracy and concentration of solutions used is rarely known with much higher precision, so making an assumption to simplify calculation of pH when $[H^+]$ is less than 5% of $C_a$ and putting

$$C_a - [H] \approx C_a$$

and above equation as

$$K_a = \frac{\left[H^+\right]^2}{C_a}$$

or just

$$\left[H^+\right] = \sqrt{K_aC_a}$$

Here similar results are obtained with two other methods when assuming that dissociation fraction is less than 5% so above equation be simplified as

$$[HA] = C_a$$

which on combining with equations gives net result shown above. Here, equation tenth can be simplified by initially considering $K_aC_a$ product in many cases larger than $K_a^{2}$ hence

$$\left[H^+\right] = \frac{K_a}{2} + \sqrt{K_aC_a}$$

Also in most cases

$$\sqrt{K_a C_a} >> \frac{K_a}{2}$$

so neglecting $K_a/2$ we get equation once again. Now compare results of pH calculation of acetic (ethanoic) acid solution with $pK_a$=4.75 using all calculation methods where numbers in parentheses are dissociation fractions calculated for pH as shown in table.

Results of pH calculation of acetic acid

| $C_a$ (M) | pH calculator | pH eq. 8.10 | pH eq. 8.13 |
|---|---|---|---|
| 1 | 2.38 (0.004) | 2.38 (0.004) | 2.37 (0.004) |
| 0.1 | 2.88 (0.013) | 2.88 (0.013) | 2.87 (0.013) |
| 0.01 | 3.38 (0.041) | 3.38 (0.041) | 3.37 (0.042) |
| $10^{-3}$ | 3.90 (0.125) | 3.90 (0.125) | 3.87 (0.133) |
| $10^{-4}$ | 4.47 (0.342) | 4.47 (0.342) | 4.37 (0.422) |
| $10^{-5}$ | 5.15 (0.714) | 5.15 (0.714) | 4.87 (1.334) |
| $10^{-6}$ | 6.02 (0.949) | 6.02 (0.949) | 5.37 (4.217) |
| $5\times10^{-7}$ | 6.30 (0.972) | 6.31 (0.980) | 5.53 (5.96) |
| $10^{-7}$ | 6.79 (0.991) | 7.00 (0.994) | 5.87 (13.335) |
| $10^{-8}$ | 6.98 (0.994) | 8.00 (0.999) | 6.37 (42.170) |

As seen, dissociation fraction of 5% (0.05) gives exact result, so using simplest equation as shown above. As the dissociation fraction results higher than 5%, then above equation will start giving incorrect results.

Now checking the results of pH calculation for other weaker acid as hydrocyanic (HCN) with $pK_a$=9.31 from the table shown below.

Results of pH calculation of hydrocyanic acid

| $C_a$ (M) | pH calculator | pH eq. 8.10 | pH eq. 8.13 |
|---|---|---|---|
| 1 | 4.66 (0.000) | 4.66 (0.000) | 4.65 (0.000) |
| 0.1 | 5.15 (0.000) | 5.16 (0.000) | 5.15 (0.000) |
| 0.01 | 5.65 (0.000) | 5.66 (0.000) | 5.65 (0.000) |
| $10^{-3}$ | 6.15 (0.001) | 6.16 (0.001) | 6.15 (0.001) |
| $10^{-4}$ | 6.62 (0.002) | 6.66 (0.002) | 6.65 (0.002) |
| $10^{-5}$ | 6.91 (0.004) | 7.16 (0.007) | 7.15 (0.007) |
| $10^{-6}$ | 6.99 (0.005) | 7.66 (0.022) | 7.65 (0.022) |
| $5\times10^{-7}$ | 6.99 (0.005) | 7.81 (0.031) | 7.80 (0.031) |
| $10^{-7}$ | 7.00 (0.005) | 8.17 (0.068) | 8.15 (0.070) |
| $10^{-8}$ | 7.00 (0.005) | 8.70 (0.198) | 8.65 (0.221) |

As dissociation fraction is always very low results from both simplified equations are almost identical - but they are wrong once the concentration of acid falls below $10^{-4}$ M when the water dissociation enters the scene. It is noted that weaker the acid, earlier it will result from simplified equations.

Now in case of weak bases, similar assumptions are applied for calculating but the only difference is that, it is more convenient to find $[OH^-]$ and pOH and converting it to pH for doing calculations for $[H^+]$ from start like:

$$\left[OH^-\right] = \sqrt{K_b C_b}$$

can be converted to

$$\left[H^+\right] = \sqrt{\frac{K_w K_{aBOH}}{C_b}}$$

# References


Anderson PW, Hasegawa H (1955) Phys Rev 100: 675–681.

Asthagiri D, Dillet V, Liu T, Noodleman L, Van Etten RL, Bashford D (2002) J Am Chem Soc 124: 10225–10235.

Asthagiri D, Liu T, Van Etten RL, Noodleman L, Bashford D (2004) J Am Chem Soc 126: 12677–12684.

Beinert H, Holm RH, Munck E (1997) Science 277: 653–659.

Brazeau BJ, Lipscomb JD (2000) Biochemistry 39: 13503–13515.

Brunold TC, Solomon EI (1999) J Am Chem Soc 121: 8277–8287.

Bull C, Fee JA (1985) J Am Chem Soc 107: 3295–3304.

Bull C, Niederhoffer EC, Yoshida T, Fee JA (1991) J Am Chem Soc 113: 4069–4076.

Chen K, Hirst J, Camba R, Bonagura CA, Stout CD, Burgess BK, Armstrong FA (2000) Nature 405: 814.

Corey, E. J., and Xue-Min Cheng. 1995. The Logic of Chemical Synthesis. New York: John Wiley. ISBN 0471509795.

Deeth RJ, Fey N (2004) J Comput Chem 25: 1840–1848 PubMed.

Dickerson LD, Sauer-Masarwa A, Herron N, Fendrick CM, Busch DH (1993) J Am Chem Soc 115: 3623–3626.

Dickerson RE, Geis I (1983) Hemoglobin: structure, function, evolution and pathology. Cummings, Menlo Park.

E. J. Corey and Xue-Min Cheng, The Logic of Chemical Synthesis (New York: John Wiley, 1999, ISBN 0471509795).

Edgecombe KE, Becke AD (1995) Chem Phys Lett 244: 427–432.

Fontecilla-Camps JC, Ragsdale SW (1999) Adv Inorg Chem 47: 283–333.

Frausto da Silva JJR, Williams RJP (1991) The biological chemistry of the elements. Clarendon, Oxford.

Gilchrist ML Jr, Ball JA, Randall DW, Britt RD (1995) Proc Natl Acad Sci 92: 9545–9549.

Gill PMW (1996) Mol Phys 89: 433–445.

Girerd JJ (1983) J Chem Phys 79: 1766–1775.

Greene SH, Richards NGJ (2004) Inorg Chem 43: 7030–7041.

Gritsenko OV, Ensing B, Schipper PRT, Baerends EJ (2000) J Phys Chem A 104: 8558–8565.

Gritsenko OV, Schipper PRT, Baerends EJ (1997) J Chem Phys 107: 5007–5015.

Groenhof AR, Swart M, Ehlers AW, Lammertsma K (2005) J Phys Chem A 109: 3411–3417 PubMed.

Han W-G, Lovell T, Noodleman L (2002) Inorg Chem 41: 205–218 PubMed.

Handy NC, Cohen AJ (2001) Mol Phys 99: 403–412.

Hellinga H (1996) Curr Opin Biotechnol 7: 437–441.

Hsieh Y, Guan Y, Tu C, Bratt PJ, Angerhofer A, Lepock JR, Hickey MJ, Tainer JA, Nick HS, Silverman DN (1998) Biochemistry 37: 4731–4739.

Hwang J, Krebs C, Huynh BH, Edmondson DE, Theil EC, Penner-Hahn JE (2000) Science 287: 122–125.

J. March and D. Smith, Advanced Organic Chemistry, 5th (New York: Wiley, 2001).

Jonkers G, de Lange CA, Noodleman L, Baerends EJ (1982) Mol Phys 46: 609–620.

K.C. Nicolaou and E.J. Sorensen, Classics in Total Synthesis (New York: VCH, 1996).

Koch W, Holthausen MC (2001) A chemist's guide to density functional theory. Wiley-VCH, Weinheim.

Konecny R, Li J, Fisher CL, Dillet V, Bashford D, Noodleman L (1999) Inorg Chem 38: 940–950.

Kozlowski PM (2001) Curr Opin Chem Biol 5: 736–743.

Lah MS, Dixon MM, Pattridge KA, Stallings WC, Fee JA, Ludwig ML (1995) Biochemistry 34: 1646–1660.

Li J, Nelson MR, Peng CY, Bashford D, Noodleman L (1998) J Phys Chem A 102: 6311–6324.

Li J, Noodleman L, Case DA (1999) In: Solomon EI, Lever ABP (ed) Inorganic electronic structure and spectroscopy. Methods, vol 1. Wiley, New York, pp 661–724.

Liu KE, Lippard SJ (1995) Adv Inorg Chem 42: 263–289.

Ludwig ML, Ballou DP, Noodleman L (2001) In: Wieghardt K, Huber R, Poulos T, Messerschmidt A (eds) Handbook of metalloproteins. Wiley, Chichester, pp 652–667.

Matthew H. Todd, Computer-aided Organic Synthesis, Chemical Society Reviews 34: 247–266.

McMurry, John. 2004. Organic Chemistry, 6th ed. Belmont, CA: Brooks/Cole. ISBN 0534420052.

McWeeny R, Sutcliffe BT (1976) Methods of molecular quantum mechanics. Academic, London.

Miller A-F, Padmakumar K, Sorkin DL, Karapetian A, Vance CK (2003) J Inorg Biochem 93: 71–83.

Miller AF, Sorkin DL (1997) Comments Mol Cell Biophys 9: 1–48.

Morales R, Chron M-H, Hudry-Clergeon G, Petillot Y, Norager S, Medina M, Frey M (1999) Biochemistry 38: 15764–15773.

Nicholls DG, Ferguson SJ (2002) Bioenergetics 3. Academic, San Diego.

Noodleman L, Baerends EJ (1984) J Am Chem Soc 106: 2316–2327.

Noodleman L, Case DA (1992) Adv Inorg Chem 38: 423–470 CrossRef.

Noodleman L, Lovell T, Han W-G, Li J, Himo F (2004) Chem Rev 104: 459–508 PubMed.

Noodleman L, Lovell T, Han W-G, Liu T, Torres RA, Himo F (2003) In: Lever AB (ed) Comprehensive coordination chemistry II, from biology to nanotechnology. Fundamentals, vol 1. Elsevier, Amsterdam, pp 491–510.

Noodleman L, Norman JG Jr (1979) J Chem Phys 70: 4903–4906.

Noodleman L, Post D, Baerends EJ (1982) Chem Phys 64: 159–166.

Ogliaro F, Cohen S, Filatov M, Harris N, Shaik S (2000) Angew Chem Int Ed Engl 39: 3851–3855.

R.B. Woodward, M.P. Cava, W.D. Ollis, W. D. A. Hunger, H.U. Daeniker, and K. Schenker, The Total Synthesis of Strychnine, Journal of the American Chemical Society 76 (18): 4749–4751.

Schueler-Furman O, Wang C, Bradley P, Misura K, Baker D (2005) Science 310: 638–642.

Schultz NE, Zhao Y, Truhlar DG (2005) J Phys Chem A 109: 11127–11143 PubMed.

Schultz NE, Zhao Y, Truhlar DG (2005) J Phys Chem A 109: 4388–4403.

Slater JC (1954) Phys Rev 91: 528.

Slater JC (1972) Adv Quantum Chem 6: 1.

Solomons, T.W. Graham, and Craig B. Fryhle. 2004. Organic Chemistry, 8th ed. Hoboken, NJ: John Wiley. ISBN 0471417998.

Stubbe J, van der Donk WA (1995) Chem Biol 2: 793–801.

Swart M, Groenhof AR, Ehlers AW, Lammertsma K (2004) J Phys Chem A 108: 5479–5483.

te Velde G, Bickelhaupt FM, Baerends EJ, Fonseca Guerra C, van Gisbergen SJA, Snijders JG, Ziegler T (2001) J Comput Chem 22: 931–967.

Torres RA, Lovell T, Noodleman L, Case DA (2003) J Am Chem Soc 125: 1923–1936.

Tschinke V, Ziegler T (1990) J Phys Chem 93: 8051.

Vogel, A.I., et al. 1996. Vogel's Textbook of Practical Organic Chemistry, 5th Edition. Prentice Hall. ISBN 0582462363.

Xu X, Goddard WA III (2004) J Chem Phys 121: 4068–4082 PubMed.

Xu X, Goddard WA III (2004) Proc Natl Acad Sci USA 101: 2673–2677 PubMed.

Yiklimaz E, Xie J, Brunold TC, Miller A-F (2002) J Am Chem Soc 124: 3482–3483.

Zener C (1951) Phys Rev 82: 403.

Zhao XG, Richardson WH, Chen J-L, Li J, Noodleman L, Tsai H-L, Hendrickson DN (1997) Inorg Chem 36: 1198–1217 PubMed.

Zumdahl, Steven S. 2005. Chemical Principles. New York, NY: Houghton Mifflin. ISBN 0618372067.

# Index

## A



## B



## C

www.ingramcontent.com/pod-product-compliance
Lightning Source LLC
LaVergne TN
LVHW011818200926
848484LV00005B/1235
* 9 7 8 1 7 8 7 1 5 0 1 2 6 *